There is no question that this is the most critical time in human history and what we do or don't do will make the difference as to whether we evolve or perish. Steve Schein is one of those people whose passion and commitment corresponds to the necessary turnings of our time. With warmth and originality, he shows us how we are the pilgrims and the parents of this emerging new era and how old formulas and stopgap solutions will no longer serve. His science with regard to sustainability is impeccable, his insights into the changing face of business management and philosophy are powerful and truth-telling. His deep exploration of the psyche and substance of 75 corporate executives provides a basis for hope in transforming leadership and thereby, in the words of the ancient myths, finding new ways of "greening" the wasteland.

Dr Jean Houston, author of 28 books on human and social development; Consultant on Human Development to the United Nations Development Programme

Thought-provoking and insightful, *A New Psychology for Sustainability Leadership* is of vital importance to the New Nature Movement.

Richard Louv, author, *The Nature Principle* and *Last Child in the Woods*

Steve Schein has provided a great service to the study and practice of sustainability management with this work. He has considered the psychological drivers – *the why* – and the leadership capacities – *the how* – that drive sustainability leaders. As a Collaborator-In-Chief myself, I found Schein's analysis enlightening. I have a greater grasp on the why and the how of my role and can thereby be more a more effective leader. I highly recommend this very readable book.

Chris Librie, Senior Director, Living Progress, Hewlett-Packard

D1132071

This book is one of the first to look at the personal motivation of sustainability professionals. It examines what Schein calls their ecological worldview, which Senge and others have referred to as the mental models that shape the way people think and act. Drawing on eco-psychology, environmental philosophy and related disciplines, including integral theory, Schein attempts to categorize various stages of consciousness about who we are and the world we live in. Schein's interviews of executives help to uncover a pattern in the evolution of thinking from anthropocentric to ecocentric. He rightly points out that seeing such a pattern, and shifting to an ecocentric worldview, will be critical for business if it is to produce meaningful sustainability outcomes.
Dr Chris Laszlo, Weatherhead School of Management at Case Western Reserve University; co-author, *Embedded Sustainability*

I've been waiting for a book like this for over ten years! One of the most crucial things missing in nearly all sustainability discussions is an understanding of how to foster the growth of ecological worldviews. Schein redresses this through his study of the most advanced forms of sustainability leadership. In the process, he makes a powerful case that such an approach is essential to the future of sustainability leadership.
Sean Esbjörn-Hargens, co-author, *Integral Ecology*; Founder and President, MetaIntegral Associates

In this refreshingly readable book, Steve Schein develops the basis for a psychology of successful sustainability leadership by engaging us with his own story of growing awareness of his direct interdependence with the environment, as well as with stories of the 75 corporate executives he interviewed.
Bill Torbert, PhD, author of *Action Inquiry: The Secret of Timely and Transforming Leadership*

Although academic in format, this book feels like a dinner party conversation that lasts late into an evening because it is so compelling. Evolving our core consciousness is an important quest for us and for our children and grandchildren. More immediately, besides business curriculum, this book has deep implications for HR, hiring and executive succession planning.

Anna Eleanor Roosevelt, President and CEO, Goodwill Industries of Northern New England

By offering a fascinating new perspective about the psychology of sustainability leadership, this book opens an entirely new type of conversation. It should be required reading for corporate executives, management educators and business students everywhere.

Joel Makower, Chairman and Executive Editor, GreenBiz Group

In this deeply thoughtful book, Steven Schein illuminates our many "ways of thinking" about sustainability. He shows how some of these mental maps allow us more motivation, more commitment, and more degrees of freedom to address challenges to our very survival than others. Marrying insights from psychology with an exploration of the thoughts and experiences of corporate sustainability leaders, Schein not only helps us identify the journey that is required of us all, but he provides the roadmap for our travels.

Mary C. Gentile, PhD, Harvard Business School; Babson College; author of *Giving Voice to Values: How to Speak Your Mind When You Know What's Right*

Much has been written and said recently about the importance of sustainability, but Steve Schein's book is particularly valuable. He identifies specific attributes of how effective leaders think, and what they are doing to make a difference. If you're looking for insights about how to move from good intentions about sustainability to effective action, this book offers real insights.
Dr Keith Melville, PhD, The Kettering Foundation, Fielding Graduate University

Pursuing the frontiers of sustainability requires transformational change—in society, our organizations and the individuals that lead them. By focusing on the psychological dimension of sustainability leadership, Schein's book fills a vital gap in the change-makers' toolkit. Given the scale and urgency of our global challenges, we desperately need bold action from our leaders. With a skilful mix of personal narrative, theoretical insights and fresh research, Schein helps us better understand what drives sustainability leaders—and what we need to do to turn the trickle of change into a flood.
Dr Wayne Visser, University of Cambridge, Founder and CEO, Kaleidoscope; author, *Sustainable Frontiers*

Helping people and organizations understand their relationship to, and dependence on, nature is critical to creating the lasting change and sustainable future we all want. In *A New Psychology for Sustainability Leadership* Steve Schein explores the important psychological connection people have to nature in hopes of sparking new conversations and research around corporate sustainability leadership.
Josh Henretig, Senior Director Energy, Environment and Cities, Microsoft Corporation

I have read hundreds of books on sustainability and written one. But this book is different from anything I have seen. Steve Schein writes from his heart with so much authenticity and conviction that I read the whole manuscript in one sitting. Absolutely unputdownable, full of practical ideas, ancient wisdom, and the latest theories and research: an all-in-one treat for anyone seriously thinking of a making a difference.

Tojo Thatchenkery, PhD, George Mason University; co-author, *Appreciative Intelligence: Seeing the Mighty Oak in the Acorn*

This book is a *tour de force*, beautifully charting a path to planet-changing leadership. If you want to deepen your impact and cultivate the capacity to lead transformational change, Schein's done the research. He guides you through the science of a stunning new psychology that can lead us out of the labyrinth and into a future we all want. This book and the worldview it heralds will deeply unlock potential for any change agent working toward a flourishing world.

Barrett C. Brown, PhD, Co-Founder, MetaIntegral Academy; author, *The Future of Leadership for Conscious Capitalism*

At a time when the revenues of some companies dwarf the GDP of nation-states, and supply chains are interwoven throughout the world, corporate responsibility has never been more important as a powerful force for our continued prosperity. Until now, no one has explored the psyche that motivates corporate leaders to protect people and the planet. Understanding these motives may very well be the key to unlocking a sustainable future for us all.

Tim Mohin, Director of Corporate Responsibility, Advanced Micro Devices; author, *Changing Business from the Inside Out: A Treehugger's Guide to Working in Corporations*

As the planet teeters on the edge of climate thresholds, CEOs demanding action have been sorely lacking. In this context, Schein asks: "What sets business leaders on fire about sustainability? And how can we spread that fire?" Our future might well hinge on the answers.

Auden Schendler, Vice President of Sustainability, Aspen Skiing Company; author of *Getting Green Done: Hard Truths from the Frontlines of the Sustainability Revolution*

With the information overload that characterizes modern society, rare is the book that stands out from the rest. This one is it. Rather than give us the usual outline of all the problems involving why human societies are not sustainable followed by the same, tired prescriptions of what we need to change, this work presents new ideas with a twist. To create change, we simply first need to be awake—in the fullest sense of that term. Second, we need to understand self and use self as the instrument for radical change. Refreshing, provocative and inspiring in turns, this book should be on every bookshelf and E-reader in the country.

Katrina S. Rogers, President, Fielding Graduate University

In exploring the deeper psychological motivations of corporate sustainability leaders, Steve Schein opens an exciting door through which so many of us will eagerly step. His ground-breaking work rightly deserves the attention of all those working towards more meaningful forms of corporate social responsibility.

Donnie Maclurcan, PhD, FRSA, Executive Director, Post Growth Institute; author of *Nanotechnology for Global Sustainability*

This easy-to-read book unveils a number of deeply held assumptions held by corporate leaders whose organizations maintain ecological sustainability priorities. Knowing them might help others create a "new" management psychology that will at long last consider the seventh generation.
Four Arrows, aka D.T. Jacobs, PhD, EdD, co-author of *Differing Worldviews in Higher Education*

Steve Schein has gotten great feedback from some powerful executives and proposes a new way of looking at sustainability in this book. If your career path is taking you in the direction of CSR or any role in the sustainability movement, I heartily recommend it as a must for your library.
John Renesch, US futurist; author of *The Conscious Organization* **and** *Conscious Leadership*; **founder, FutureShapers**

Brilliant. Schein's work provides a look into a gaping hole in the sustainability literature. His research reveals the essence of sustainability leadership (and human behaviour in general): how we *see the world* informs how we think, understand and act. This engaging and provocative book examines the psychological underpinnings of an ecological worldview and leads the reader to a new understanding of sustainability leadership.
Mary A. Ferdig, PhD, President and CEO, Sustainability Leadership Institute

Along with technological innovation, harnessing market forces and transparency in the supply chain, there is an inner landscape of values and mental maps that guide behaviour in would-be green businesses and organizations. In *A New Psychology for Sustainability Leadership*, Steve Schein helps to reveal the important and often overlooked "ecology of mind" that distinguishes effective sustainability leadership and performance. I highly recommend it.
Thomas Doherty, PsyD, Founding-Editor, *Ecopsychology Journal*; Past President, Society of Environmental, Population and Conservation Psychology

Confronting sustainability is about grappling with complexity in our social as well as natural environments. This book explores how leaders who surround themselves with that complexity end up rising to the challenge by growing their own cognitive and emotional complexity. While this central insight is an important intellectual contribution, the value of this book is its weaving of personal reflection and narrative with the journey of the research.
Jason Jay, PhD, Senior Lecturer and Director, Sustainability Initiative, MIT Sloan

In this well-researched book, Schein shines a light on a poorly considered yet key aspect of sustainability leaders: the worldviews shaping their actions. The good news is that we have available the greatest tools for this task, and they are in the hidden power of the ecological worldview. Something to relearn that can mean the difference between our civilization's breakdown and breakthrough. And it's at our fingertips.
Isabel Rimanoczy, Scholar in Residence at Nova Southeastern University; author of *Big Bang Being*

Schein has written a compelling book that helps us understand the values and motivators that drive sustainability leaders. He has not only provided a framework for building leaders who can drive sustainability in the corporate world, but for awakening each of our ecological selves so that we understand our part in sustainability. What I find even more enthralling is that Schein writes as if he was sitting by a campfire, telling a story. An enlightening book which I highly recommend.

Richmond S. Fourmy, PsyD, Associate Partner, Aon Hewitt

Schein promotes the message that continued evolution of thought and human development requires enhanced awareness of self, others and the environment in which we operate. This awareness is captured in the concept of ecological self. He emphasizes the opportunity to further develop current and future leaders of global organizations to make more complex, humane and eco-sensitive decisions that embrace a new psychology of sustainability leadership.

John Bowling, PhD, Managing Partner, Sustainable Leadership Consultants

Readers will sail through this easy-to-understand, well-written and timely book on the complex and important subject of worldview among current and future corporate leadership for sustainability.

Kerul Kassel, PhD, author, *The Thinking Executive's Guide to Sustainability*

Steve Schein is the rare individual who can take on the dual perspective of manager and scholar to develop innovative yet practical lessons for business professionals.

Elliot Maltz, PhD, Professor, Atkinson Graduate School of Management, Willamette University

This book provides hope and inspiration to those of us worried that we are not getting the sustainability leadership we need. Schein provides sophisticated insight into the psychological evolution that must take place within the corporate world and evidence that it is happening amongst the leading edge of corporate sustainability leaders. Drawing on an immense interdisciplinary literature, his direct experience in the corporate world, his more recent teaching experiences, his ecological self, and dozens of interviews with corporate leaders, Schein has produced a book that should find its way onto course syllabi at business schools, environmental studies and boardrooms.
Paul Haber, PhD, Professor of Political Science, University of Montana

Just as corporations must make changes in their cultures, corporate leaders must make changes in their thinking. This book offers a fresh perspective and offers new insights about how companies can advance and improve their efforts to become more sustainable.
Andrew Savitz, author, *The Triple Bottom Line*

A New Psychology for Sustainability Leadership

The Hidden Power of Ecological Worldviews

A NEW PSYCHOLOGY
FOR SUSTAINABILITY LEADERSHIP
The Hidden Power of Ecological Worldviews

STEVE SCHEIN, PhD

Greenleaf
PUBLISHING

© 2015 Steven Schein

Published by Greenleaf Publishing Limited
Aizlewood's Mill
Nursery Street
Sheffield S3 8GG
UK
www.greenleaf-publishing.com

Cover by Sadie Gornall-Jones

Printed and bound by Printondemand-worldwide.com, UK

British Library Cataloguing in Publication Data:
    A catalogue record for this book is available from the British Library.

    ISBN-13: 978-1-78353-190-5 [hardback]
    ISBN-13: 978-1-78353-195-0 [paperback]
    ISBN-13: 978-1-78353-191-2 [PDF ebook]
    ISBN-13: 978-1-78353-192-9 [ePub ebook]

We are in a horse race with catastrophe. Can corporations move fast enough? Government cannot. It will not. Corporations might. Will they? I don't know. And that turns the future of the world.

Hunter Lovins

Contents

Part 1: Introduction

Part 2: Exploring the corporate eco-psyche

Part 3: How sustainability leaders think

Part 4: The future of sustainability leadership

Figure

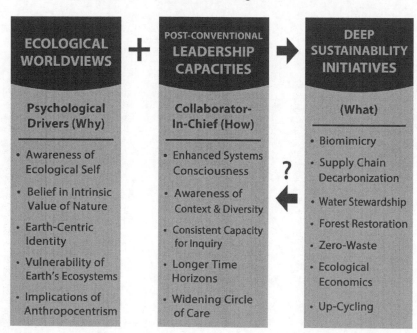

The hidden power of ecological worldviews

ECOLOGICAL WORLDVIEWS	POST-CONVENTIONAL LEADERSHIP CAPACITIES	DEEP SUSTAINABILITY INITIATIVES
Psychological Drivers (Why)	Collaborator-In-Chief (How)	(What)
• Awareness of Ecological Self	• Enhanced Systems Consciousness	• Biomimicry
• Belief in Intrinsic Value of Nature	• Awareness of Context & Diversity	• Supply Chain Decarbonization
• Earth-Centric Identity	• Consistent Capacity for Inquiry	• Water Stewardship
• Vulnerability of Earth's Ecosystems	• Longer Time Horizons	• Forest Restoration
• Implications of Anthropocentrism	• Widening Circle of Care	• Zero-Waste
		• Ecological Economics
		• Up-Cycling

Tables

Preface

Today, there are major changes happening inside a select group of the world's largest corporations. In response to the Earth's most pressing environmental problems, a growing number of the most recognizable multinational companies are transforming the way they do business. These new developments hold great potential for our future.

Using a process known as biomimicry, engineers are designing new products based on a more thorough understanding of how nature works. Using a tool known as life-cycle analysis, accountants are measuring the full environmental footprint of products, from resource extraction through the manufacturing supply chain, distribution, disposal and recovery. Large-scale collaborative efforts between multinational companies, environmental nonprofits and governments are leading to new systemic approaches to our most complex global environmental problems involving the oceans, farmlands, forests, river basins and our fellow species.

However, we all know this has not been close to enough. Whether as sustainability educators, CSR executives, environmental activists or simply concerned citizens, we worry about each new piece of depressing environmental news. For those of us that follow the corporate sustainability movement, we know that only a relatively small percentage of progressive corporations are thoroughly integrating sustainability initiatives throughout their global organizations. We also know that only certain executives within these corporations are fully committed to sustainability as their highest priority. As a result, the quarterly earnings report is still the major driver in the corporate world and CEOs with sustainability at the top of their agenda are few and far between.

A focus on corporate sustainability executives

During the last decade, the sustainability position in multinational corporations has grown considerably in influence. Beginning with the appointment of the first chief sustainability officer in 2004,[1] today there are senior sustainability executives in hundreds of the world's largest multinational companies. In many cases, the chief sustainability officer now reports directly to the CEO. These are highly influential individuals inside today's global corporations.

Behind each major environmental announcement by a multinational CEO, a small group of executives dedicate themselves to a wide range of sustainability initiatives, much of it in the face of strong resistance. Their companies have tens of thousands of employees throughout the world. Their global supply chains affect millions of people. Their customers reach into the billions.

On the one hand, we can blame corporations as a whole for the ecological crisis. However, when we consider their potential to radically reduce their impacts, reinvent their energy sources and repurpose their infrastructure to eventually restore Earth's ecosystems, the sustainable business movement may be the single most important environmental movement in the world today. When we take into account how this affects the availability of food and water for the planet's poorest people, a case can likewise be made that it is also the most important social justice movement.

> The sustainable business movement may be the single most important environmental movement in the world today

Despite the global scale of their companies, the number of executives who champion sustainability initiatives on a daily basis is surprisingly small. Although much has been written about their accomplishments, we don't know enough about their personal histories, their deeper motivations and how they think. We don't know enough about how they think about nature, leadership, resistance and change. At its essence, we don't know enough about what makes these types of global sustainability leader tick.

The limits of "sustainability"

At the same time, we have become aware of the limits of the term "sustainability". We know that it can mean very different things to different people. We have seen how it can be used narrowly to mean short-term economics, jobs and national security; or it can used expansively to mean a complete transformation to deeply ecological and restorative business models. Sadly, we have witnessed how sustainability can be misused and misunderstood. In our darkest moments, we fear that the sustainability movement has fallen well short of its overall goal to transform business and society.

We continue to ask ourselves, why? Why, despite all the scientific evidence, don't all senior executives have a strong sense of urgency about transforming business in response to climate change? Why doesn't everyone see the clear and deep connections between our traditional ways of doing business and harming the ecosystems we depend on for life? Why is there so much resistance to change? Although we tell ourselves that politics, jobs and our fossil-fuel-dependent economy and culture are the obvious reasons, we continue to search for new answers.

I've written this book to offer a new type of answer to these questions and a new place to look for solutions.

Cultivating a new psychology for sustainability

Lester Brown, President of the Earth Policy Institute and author of more than 50 books on global environmental issues, observes, "Every political movement has its psychological dimension. Persuading people to alter their behavior always involves probing motivations; activism begins with asking what makes people tick? The environmental movement is no exception."[2]

Although thousands of sustainability-related books, articles and corporate reports have been published in recent years, today little is known about the deeper psychological motivations of corporate sustainability leaders. Leadership consultant and human development researcher Barrett C. Brown observes that the more we understand how psychology and worldviews drive the behaviours required to lead sustainability initiatives, the more effectively we will be able to cultivate them, especially during times of complexity and rapid change.[3]

> Little is known about the deeper psychological motivations of corporate sustainability leaders

As sustainability educators, executives and activists, we need to develop a new, shared understanding of what sustainability leadership must become. We need a new story, a new language and, most of all, a new psychology.

Ecological worldviews: a missing perspective

The research I share with you in this book draws on eight distinct social science traditions that have not been widely used to study corporate sustainability leadership. These include eco-psychology, deep ecology, ecological economics, social psychology, environmental sociology, indigenous studies and the new field of integral ecology. I also rely on developmental psychology research about how worldviews are constructed, how we interpret the world around us, and how this can change over the course of our lives.

Using key insights from these disciplines, I include in the book extensive quotations from my interviews with 75 global sustainability leaders in more than 40 multinational organizations. The interviews suggest that many of the most influential corporate sustainability leaders are motivated by their ecological worldviews, which can be thought of as the deep mental patterns and ways of seeing our relationship to the natural world. Ecological worldviews can also be thought of as our cognitive and perceptual capacity to see the world through the lens of ecology, which is essentially the relationship of species and their environment.

> Many of the most influential corporate sustainability leaders are motivated by their ecological worldviews

In the minds of sustainability leaders, ecological worldviews can enhance the perception of our interdependence with the Earth's planetary ecosystems, which can strengthen the depth of their commitment in the face of continued resistance. The interviews further reveal expressions of what developmental psychologists call post-conventional worldviews, which can enhance their ability to effectively communicate to diverse audiences, collaborate across boundaries and unlock capacity to lead large-scale transformational change.

MIT professor emeritus and long-time sustainability scholar John Ehrenfeld reflects that, in order to address sustainability fully and meaningfully, we must make fundamental shifts in the way we think. Referring to our capacity to lead transformational change, he invites us to consider that, in the face of opposition, an individual can always change his or her own worldview.[4]

> For too long we have assumed that all multinational corporations, and by default all executives inside them, have the same worldview

For too long we have assumed that all multinational corporations, and by default all executives inside them, have the same worldview. If we are to advance the field of sustainability leadership beyond its current limitations, it is vital to understand how global sustainability leaders think, how their worldviews have been formed, and how this influences their actions.

By shining a light on the psychological dimensions of a large group of sustainability executives in multinational corporations, my hope is this book will open up new conversations and new research across a wide range of social science disciplines in the context of corporate sustainability leadership. Ultimately, this can lead to a new psychology for sustainability that can be integrated into public and private institutions everywhere to support the development of the next generation of sustainability leaders for the benefit of all life on Earth.

Acknowledgements

Towards the end of the dissertation research that led to this book, my wife and I travelled to India and Nepal to visit our daughter, who was studying in Kathmandu. One morning before dawn, we found ourselves on a small rowboat on the Ganges River in the ancient city of Varanasi.

As the sun rose along the eastern shore, we placed floating candles on the river, praying and chanting with our Hindu guide. At one point, I turned to my wife and said, "I feel as though my dissertation committee is along with me." Although she had become accustomed to how often I brought them up during the previous year, I don't think she expected me to mention them at precisely that moment! When she asked me what I meant, I explained how they were each connected to India, and to me, in different ways.

My chair, Fielding President Dr Katrina Rogers, was about to give a keynote address on ecological leadership at a management conference in New Delhi. Anthropologist

Dr David Willis, my first faculty reader, was conducting research on a spiritual community in Southern India. My second faculty reader, human development scholar Dr Judy Stevens-Long, had recently begun a new book about several of the world's most important religious texts, including the Bhagavad Gita. My external examiner, MIT sustainability scholar Dr Jason Jay, just celebrated the birth of his daughter Uma, which means mother goddess Durga in Hindu mythology. My student reader, Julie Huffaker, travelled and meditated in India years ago and had shared with me how deeply the trip had affected her life.

I mention these connections to India for several reasons. First, they offer insight into the interdisciplinary breadth and depth of the five individuals that comprised my committee. All of these individuals are not only highly respected scholars in multiple disciplines that include hermeneutics, cultural anthropology, human development, leadership, organization development and sustainability, but they are also great teachers.

However, this is not why I was thinking about them that morning on the Ganges. I was thinking how each of them helped me *see* and *feel* so much more of the human story of this ancient civilization while floating down the river. I realized at that moment that my doctoral studies had given me much more than a PhD, they had greatly enhanced the way I will experience the rest of my life. At 54 years of age, this is no small gift. I am very grateful for the guidance of these five individuals during my dissertation research.

I thank the many Fielding faculty members from whom I learned much including Dr Don Trent Jacobs (aka Four

Arrows), Dr Richard Appelbaum, Dr Keith Melville, Dr Tojo Thatchenkerry, Dr Anabelle Nelson and Dr Charles McClintock.

I also express my gratitude to several scholars outside of Fielding with whom I developed friendships while conducting this research. The many conversations I had with each of them at pivotal moments helped me back out of more than a few corners and open new doors: Drs John Bowling, Barrett C. Brown, Richmond Fourmy, Sean Esbjörn-Hargens, Annick Hedlund-de Witt, Hilary Bradbury Huang, Elliot Maltz, Aliki Nicolaides, Paul Haber, William Torbert and Nancy Wallis.

I thank Becky Macklin, Dean Bargh and the team at Greenleaf Publishing for their interest in my research and their support in expanding my dissertation into this book. I thank the 75 sustainability leaders that took the time to be participants in my research as well as the many sustainability leadership students I've had over ten years teaching at Southern Oregon University. I also thank the Carpenter Foundation and SOU for their support of my research.

I am deeply grateful for the circle of close friends that took walks in the woods and listened patiently as I worked out ideas including Ben Bellinson (and his giant dog Nelson), Dana Carmen, Ken Crocker, Aaren Glover, Matt Hough, Jacob Kahn, Stephen and Sarah Marshank, Eric Ring, Steve Scholl, Stephen Sendar and Robert Townsend.

I express gratitude to my family of origin, Gloria, Theodore, Peter and Amy, who shaped my worldview early in life. Finally, I offer a deep bow to my family. My three amazing children: Casey, Maggie and Ted, ages 24, 22 and

20; each of them stepping into their own worldviews in such wonderful ways these days. My partner of almost 30 years, Patty Samera Schein, thank you for being alongside me, behind me and putting up with me, while I wrote this book.

Part 1: Introduction

Sustainability is both a badly misused and abused
term.

John Ehrenfeld

1
Ecologically awake

The roots of this book are in the garden. It was there, ten summers ago, where the fragility of the Earth's ecosystems we depend on for life first became clear. At the age of 45, this realization, and its implications for business, came as a shock.

It started in the spring of 2006, when I became steward of a ten-acre piece of land in the hills above the town of Ashland, Oregon. The land was north facing and densely forested with a mix of ponderosa trees, black oaks, Douglas firs, and western red cedars. Underneath the canopy of trees grew a thick carpet of native grasses, manzanita and wild rosebushes.

The mosaic of the forest made me deeply curious about the landscape around me. What was the history of this land? Why were certain trees healthier than others? Where was water flowing beneath the forest floor? How was the microclimate changing? These types of questions led me to a series of environmental science teachers that started with an off-the-grid ethno-botanist and permaculture[5] teacher named Tom Ward. I first met Tom in the old oak grove at the bottom of my property. He wasted no time in beginning my full-immersion course of applied environmental studies.

He started with fire ecology and ethno-botany, the study of relationships between human beings and plants, a field that can be traced back to Pythagoras in the sixth century BC. In the weeks that followed, Tom taught me how to create a perennial food forest and how to build topsoil through cover crops and composting. He showed me how to bring the bees through continuous pollination and support beneficial insects through discontinuous planting patterns.

Tom explained how perennial polyculture and biodiversity create resilience. He described closed-loop systems, waste repurposing, and how to catch rainwater by building bio-swales and small ponds across the land. He explained the historical role of fire in the ecology of the forest. More than through his words, Tom powerfully role modelled how to observe nature closely every day and live more lightly on the Earth.

Later that winter, when the fall rains softened the clay baked hard by the summer heat, we began to survey and dig a series of trails snaking up the hillside through the trees. We used a large supply of partially burnt logs found scattered in the forest to create earthen berms on the downhill side of the trails, using the wet clay to sculpt the trails like cement provided courtesy of nature.

In spring we dug postholes and built a deer fence around one acre for the garden. We created a network of drainage swales and ponds designed to hold water along the hillside and down in the garden during the rainy season. With the help of several friends, we thinned the densely covered forest, felling hundreds of small diameter ponderosa trees, using draw-knives to peel the bark, and piling the beautiful yellow logs in crisscrossed log decks that resembled rafts.

The following summer we dug test pits and took soil samples of clay and sandstone. We unearthed dozens of large boulders and slabs of sandstone covered with shell fossils, which came from an ancient geological formation when the entire area was covered by ocean.

We found many uses for these stones, from foundations for a greenhouse in the garden, to retaining walls and stepping stones. These fossilized rocks spoke of a time when the valley was covered by sea. They provided hard evidence of geological time.

A big part of those first summers on the land involved learning how to grow food based on principles of permaculture and organic farming. This led to a realization about our industrial food system and the inescapable environmental implications for the planet. Along with Tom's hands-on permaculture teachings, there were several books and articles that completely changed the way I thought about food. First, *The Omnivore's Dilemma* by food writer Michael Pollan explained how our industrial food system was leading to massive soil erosion, vanishing species, the obesity epidemic and a host of other health and economic problems.[6]

Next, *Animal, Vegetable, Miracle* by novelist Barbara Kingsolver and her husband, Steven Hopp, an environmental studies professor, described the full long-term implications of large-scale industrial, genetically modified and petroleum-driven agriculture. They explained how, through the oil and natural gas consumed by the machinery, fertilizers, pesticides, processing and transportation of large-scale corporate agriculture, we Americans put almost as much oil into our stomachs as into our cars. In fact, Americans consume more than 400 gallons of oil per year per citizen, or almost 20% of our nation's energy use.[7]

Most importantly, a global systems perspective of the Earth's ecological crisis was provided in *Plan B*, the comprehensive assessment of the Earth's natural resources by Lester Brown, President of the Earth Policy Institute.[8] As one of the world's most influential environmental thinkers, Brown has been synthesizing and communicating the large-scale environmental trends and their implications throughout the world for more than 50 years. In *Plan B*, Brown and his team of researchers present the environmental data, ecosystem by ecosystem, from rising temperatures and sea levels, to melting ice caps and deforestation, to eroding soil and falling water tables. Most disturbingly, they describe how growing food and water shortages are contributing to the increasing political instability we're seeing around the world.

A few summers later, a large wildfire broke out less than a half-mile east of my land. As my friend Aaren and I raced around the perimeter of the house cutting down dozens of ponderosa trees, helicopters and planes loaded with fire retardant flew directly over our heads making laps, dropping their loads on the fire. My adrenalin was flowing, my heart was racing, and large clouds of billowing black smoke filled the sky and blocked out the sun, adding to the surreal scene.

The wildfire turned out to be the last of that summer's fire season. While it did burn over 200 acres including one occupied home, it never crossed the road to directly threaten our house. However, in its aftermath, something interesting occurred in the forest at the top of the hill on our land.

The following winter, the evergreen needles of hundreds of ponderosa trees turned brown all at once and started dying from the top down. A local arborist helped me identify this phenomenon as the handiwork of the California pine beetle, otherwise known as *Ips paraconfusus*.[9] Evidently, large swarms of these beetles were flushed out of the forest across the road during the previous summer's forest fire. They then flew west and landed in the trees on my land. Once there, they began to bore holes in the bark, a process that eventually strangles each tree and prevents it from being able to suck water up through its roots to the crown.

These experiences allowed me to comprehend more of the complexity of the natural environment that surrounds us, what physicist and systems thinker Fritjof Capra calls the "web of life".[10] It made me more aware of how much I had taken for granted about Earth's ecosystems and how far out of balance we had become as a global society. It gave me a glimpse of what changes in these systems could mean for the future of this land and, on a much, much larger scale, for the future of all people.

> These experiences made me more aware of how much I had taken for granted about Earth's ecosystems

It also made me painfully aware of how much I had missed during my own business school education and during my subsequent corporate career as the president and co-founder of several companies, the last of which became a public company in 1998.

In 2005, I left the corporate world to join the faculty of a university business school. This career shift, which coincided with my move to the land, provided me with an opportunity to apply ecological awareness to business education.

1.1 A higher purpose for business

During the first few years I taught courses that were part of the standard curriculum. These included courses on strategic management, organizational leadership and business ethics. However, in the fall of 2007 a pivotal conversation with a friend led me to attend the Bioneers conference in San Rafael, California. Founded more than 25 years ago by Kenny Ausubel and Nina Simmons, Bioneers is one of the first annual conferences that bring together environmental, social justice and corporate responsibility activists under one roof. Based on principles of biological and cultural diversity and biomimicry, the annual event has inspired an entire generation of sustainability leaders.[11]

My experience at the conference led to a second fundamental truth about sustainability: the interdependence between big business, healthy ecosystems and basic human rights. This is especially important to understand in regards to multinational corporations and their environmental impacts on developing countries around the world where much of their raw materials are sourced and their products are manufactured.

As a recent graduate from the Presidio Green MBA programme in San Francisco, my friend had his finger on the pulse of the latest sustainability trends. When he invited me I was in the middle of teaching three fall semester classes, and attending this conference had not been in my plans. It was also my eldest daughter Casey's 17th birthday and we had plans to celebrate at home with our family. However, when I told her about the conference, she looked me in the eye and said, "You have to attend this conference, and I'll come with you!"

From the first presentation I found myself thinking, if only this information could be a standard part of the business curriculum, if only these types of presentations could be offered on a broader scale at business schools, if only …

At Bioneers, I listened to paradigm-shifting talks about biomimicry by Janine Benyus, ecological literacy by David Orr, natural capitalism by Paul Hawken, the mycelial web by Paul Stamets, food policy by Michael Pollan, and indigenous rights by Agnes Baker Pilgrim and the 13 Indigenous Grandmothers.

However, there was one individual who I met at the conference who made the biggest impression on me. His name

> At Bioneers, I listened to paradigm-shifting talks about biomimicry, ecological literacy, natural capitalism, the mycelial web, food policy and indigenous rights

was Ray Anderson and, until his death in 2011, he was the CEO of Interface, one of the world's largest manufacturers of commercial modular carpet. He was the first CEO of a multinational public company to put a stake in the ground for a zero ecological footprint and adopt a "do no harm to the Earth" policy in a large-scale industrial company. Since the mid-1990s, he had been working tirelessly to help the business world and corporate leaders everywhere fully understand the environmental realities of our planet and the responsibility of big business.

In thousands of speeches, Anderson continually pointed out that business is the most pervasive and powerful institution on the planet and responsible for most of the damage done to the Earth's ecosystems.

He made a powerful argument that business must take the lead towards sustainability and restoration. Anderson also highlighted that one of the biggest things that need to change is the education system, and observed that universities everywhere are still teaching a system that is destroying the biosphere.[12]

Ray Anderson acted as a powerful role model for what sustainability leadership looks like at the CEO level, by articulating the fundamental responsibility of multinational corporations to lead the way in a transformation of our globalized economic system towards more sustainable world. Beginning in 1994, his company pioneered many of the innovative sustainability strategies

> Ray Anderson acted as a powerful role model by articulating the fundamental responsibility of multinational corporations

that were to become more common in the two decades that followed. He was guided by a deep sense that there is no bigger issue for a company than its ultimate purpose and challenged us to understand that business must exist for a higher and nobler purpose than just making a profit.

Much has been written about Ray Anderson and I won't say more here, other than that the short conversation I had with him seven years ago led to the next epiphany for me. I grasped the truth of the ecological crisis and the implications for business. It was at that moment that I decided to join the sustainability movement dedicated to transforming business in service of social and ecological justice.

1.2 Sustainability curriculum 1.0

After returning from the Bioneers conference, I started a new sustainability leadership programme at the university. On one level, my idea was simple: I wanted to get environmental studies and business students in the same room. I felt that by creating new opportunities for students from these two separate schools to come together more frequently, they could greatly enhance each other's education. However, based on the long tradition of organizing universities into separate disciplinary silos, students from these two schools did not have many opportunities to study together. I believed

> My idea was simple: I wanted to get environmental studies and business students in the same room

strongly that, in order to prepare the next generation of business leaders for future ecological challenges, they needed to have more environmental science in their curriculum.

Using principles of permaculture and natural capitalism as an initial teaching framework, I created two new sustainability-related courses for business students. Closed loops, service and flow, radical resource productivity, natural building, resilience, repurposing waste, and restoration—all principles of permaculture and natural capitalism— translated into the many ways that sustainability was beginning to be put into practice by multinational corporations.

> Principles of permaculture and natural capitalism translated into the many ways that sustainability was put into practice by multinational corporations

The first course focused on the latest sustainability innovations at multinational corporations and was designed to give students a broad overview of the applied areas of sustainability, including alternative transportation, waste reduction, renewable energy, green building, fair trade, life-cycle analysis, water stewardship, carbon footprints and other emerging sustainable business practices. As part of the class we covered more than two dozen corporate sustainability case studies, including Interface, Patagonia, Ben and Jerry's, Stonyfield Farms, Seventh Generation, TerraCycle, Nike, Starbucks, Unilever and, to the ongoing surprise of many students, even Coca-Cola and Walmart.

The second course, "Sustainability Leadership", used the same applied areas, but focused on sustainability at the individual level. Each week, I brought in inspiring guest speakers and facilitated round table discussions with a growing circle of highly knowledgeable teachers and sustainability practitioners from around the Pacific Northwest.

The cornerstone of the course featured weekly personal sustainability projects (PSPs) where students explored sustainability from a different perspective to enhance their awareness of their personal consumer choices and associated ecological footprints. These perspectives included personal transportation, household energy, water and waste management, local organic food, consumer electronics and a variety of other personal practices. Students often experienced "aha" moments through these experiences and I became increasingly aware of the deeper shifts occurring in their minds.

My experience teaching sustainability during the first few years was filled with optimism. Students calculated their carbon footprints and we printed their numbers on the back of T-shirts that we wore during public presentations. Several inspired students created a business plan for a new organic farm and sustainability centre on our campus. They won a small grant in a state-wide competition and even received their award from Nobel Prize winner, Muhammad Yunus. A few years later their vision became a reality on several acres of campus land.[13]

On the national front, President Obama had been elected in the fall of 2008 and passed two new executive orders requiring comprehensive sustainability plans that included a wide range of environmental initiatives to be measured at all federal agencies. He appointed a new green job czar. In another applied project, my students created a green job database that ended up consisting of more than a hundred promising new sustainability-related career opportunities. Green business was on the cover of *Time* magazine. It even appeared that national climate change and energy policy was making progress in Congress. On a global level, expectations were high that at the Copenhagen summit in 2009 a worldwide climate change treaty would be adopted. The sustainability revolution was gaining momentum.

During the first summer on the land, my son Ted helped me build a deck high up in the woods in a natural clearing we came to call the middle camp. The following spring, it became an outdoor classroom for students from the university. In addition to business and environmental studies students from the sustainability leadership class, students from the Native American studies classes joined us around the fire. MBA, undergraduate business and environmental studies students, inspired by Native American teachings, passed the talking stick and shared their aspirations for business and society. It was a place where we could sit around the fire and imagine a new future based on principles of sustainability.

> In addition to students from the sustainability leadership class, students from the Native American studies classes joined us around the fire

2
The limits of "sustainability"

A series of events in 2009 began to chip away at the
unbridled optimism I felt during the preceding two years
about the sustainability movement and its potential for
transformative change. These events occurred at both
the local level of the university and on the global stage
in Washington, DC and Copenhagen, Denmark. They
also allowed me to understand better that the term
"sustainability" meant very different things to different
people.

It started during a talk I gave about corporate
sustainability to several hundred students at the university
on Earth Day, 2009. The preceding speaker, a professor of
hydrology, had given a provocative talk about water issues
around the world. He caught everyone's attention by walking
around the room and squirting people in their face with
a syringe full of water! He then offered his definition of
sustainable water as the right *amount* of water, at the right
place, at the right *quality*, and at the right *time*. He focused

on severe fresh water shortages in several of the poorest countries and how global climate change would only make this problem worse in the coming years.

Picking up on his water theme at what I thought was the *right time*, I started by describing how global food and beverage companies were changing the way they did business to reduce their use of fresh water and restore major watersheds around the world. I described several of the major changes they were making at major manufacturing plants and through their global supply chains, changes that were affecting millions of people and hundreds of rivers. I showed a few slides about the goals and metrics they were using to measure their progress in specific regions and countries.

As a former international auditor, CPA and senior executive, I knew first-hand that these new initiatives represented major change by multinationals. I had been poring over corporate sustainability reports and teaching sustainability leadership for two years at that point. Through the use of new tools such as life-cycle analysis and ecological footprints, I wanted students to know that companies today were developing a much more thorough understanding of their ecological impacts on global ecosystems. I saw all this is as a step in the right direction and a hopeful sign.

"You don't know what these corporations have done! They are destroying the Earth!" shouted a student from the back of the room

However, after about 15 minutes into my talk, I heard a distinct hiss coming from somewhere in the audience. "You don't know what these corporations

have done! They are destroying the Earth!" shouted a
student from the back of the room. This cry came from
an environmental studies student who was expressing an
obvious concern shared by many students, environmentalists
and people everywhere. She challenged me that sustainability
was all just a marketing ploy and that corporations in
general did not truly care about the environment; that it was
all about the bottom line and always would be.

So I invited her, and all the other students in the room, to
take a closer look. In addition to all the new data and specific
measurements appearing on corporate sustainability reports,
I suggested that all multinational corporations, and all senior
executives inside them, were not the same. I made a few
comparisons between consumer food companies, defence
contractors and petroleum extraction companies.

I suggested that environmental attitudes and commitment
to ecological issues varied greatly among senior corporate
executives and corporate cultures. I
suspected the motivations and personal
stories of individual executives were
more complex. However, I could not cite
any published research or specific data
to support my suspicion.

> I suspected the
> motivations and
> personal stories of
> individual executives
> were more complex.

Although much had been written
about how multinational corporations
were beginning to integrate sustainability into their
operations, business and sustainability research had
focused almost entirely on best green business practices and
environmental and social impacts. As a result, not much
information could be found about the personal backgrounds,

values or deeper motivations of corporate sustainability leaders. I also could not offer any first-hand evidence. I had not personally met or spoken with any senior executives behind the sustainability initiatives at multinational companies at that point.

2.1 Running into resistance

In the fall of 2009, a new dean arrived at the School of Business. The previous dean, who had left to become the president of another university, had been very supportive of the sustainability agenda. He shared my beliefs that business students needed a much stronger environmental science background as part of the curriculum and had been a strong ally. He had wholeheartedly approved of my initiative to create a new sustainability leadership programme and helped me usher the programme through the University Curriculum Committee, Faculty Senate, and through the State Board of Higher Education, all in a few short months. I had not realized how fortunate I had been.

The incoming dean had a very different understanding of sustainability. Not only did he have little understanding of environmental issues including the science of climate change, he appeared openly hostile towards what he called my "tree-hugger" agenda. At the university, he reduced sustainability to be only about short-term financial issues for our department. Despite the evidence I provided to him about new ways that sustainability was being adopted in the corporate world, he was cynical and often sarcastic. I remember a particularly discouraging conversation where he perceived sustainability to be just "another marketing fad" that would come and go. Although this individual departed a few years later, these types of conversations and experiences made me acutely aware of the limits of the term "sustainability" and how it could be misinterpreted, and in some cases abused.[14]

> These types of conversation and experience made me acutely aware of the limits of the term "sustainability" and how it could be misinterpreted, and in some cases abused

On the world stage, as we all know, the 2009 Copenhagen Climate Summit failed to produce a global treaty to reduce carbon emissions. During the following year, climate and energy legislation was abandoned in the United States Congress. The new green jobs czar, Van Jones, was even forced to resign without being replaced.

During the subsequent two years I gave numerous lightly attended public presentations in our region and at the university. Almost always the majority of those present were environmental studies students or local citizen activists. At other sustainability-related events around the university hosted by environmental studies or Native American studies programmes, I rarely saw members of the business faculty or business community. At one faculty retreat I shared an abstract entitled "The emergence of the chief sustainability officer: A new archetype in corporate leadership". It appeared to generate little interest.

I began to wonder why, despite all the scientific evidence about the ecological crisis, and all the ways that multinational corporations were changing, my dedicated colleagues—who were highly qualified as business educators—did not appear to feel the same urgency and relevance about climate change and sustainability.

Although students that had come through my classes appeared to be readily embracing sustainability concepts and many were implementing them in their career plans and personal lives, it was clear that I was having little success at changing the overall focus of the business curriculum or the overall short-term economic mind-set of the local business community. I began to see this

> Most of all, I became aware of the limits of my effectiveness to drive change

as symbolic of both higher education and the business world in general. It allowed me to appreciate the type of resistance sustainability activists face in all types of organization. Most of all, I became aware of the limits of my effectiveness to drive change.

As long-time CSR executive and author Tim Mohin describes, "While rewarding on many levels, the job of the corporate treehugger can be frustrating ... At times you may feel like a metronome—vacillating between the euphoria stemming from your laudable accomplishments and the dejection from the feeling that your role in the corporate power structure sits somewhere between superfluous overhead and oblivion."[15]

I began to comprehend more fully how the 20th-century short-term economic model was still the major driver at universities and businesses everywhere. I became more aware of the bigger challenges and the limits of sustainability to lead to deeper transformational change. Reflecting on the entrenched resistance, I developed a case of sustainability fatigue.[16]

> Reflecting on the entrenched resistance, I developed a case of sustainability fatigue

2.2 The psychological roots

I began losing sleep and could not stop asking myself questions. Why doesn't everyone feel as strongly as I do about sustainability? Why, despite all the scientific evidence, doesn't everyone in leadership positions have the same sense of urgency about transforming business in response to climate change? Why doesn't everyone see the deep connections between environmental, social and economic issues?

While I knew that politics, short-term economics and fossil-fuel-dependent culture were among the obvious answers, my inquiry into these questions took an unexpected turn in the late summer of 2010.

I was on a family vacation to Salt Spring Island in British Columbia, when I picked up a pair of books in the little bookstore in the port when we arrived. We had rented a small cabin on a lake in the centre of the island. There was no Internet connection and we spent the week reading and walking. Away from daily distractions, my mind relaxed and was receptive to new ideas.

The books were both anthologies, *The Deep Ecology Movement*[17] and *Ecopsychology: Restoring the Earth, Healing the Mind*.[18] They contained a series of short essays by writers including Norwegian philosopher Arne Naess, human ecologist Paul Shepard and deep ecologist Joanna Macy, among many others. Collectively, these essays served as my introduction to environmental philosophy and psychology. They presented a set of ideas and concepts that were entirely new to me. These included the **ecological self**, a **new ecological paradigm**, and the distinction between **ecocentric** and **anthropocentric** worldviews.

The combination of these ideas sparked a new set of deeper questions for me about sustainability, about corporate executives and about myself. To what degree did I experience myself as literally being part of nature? To what extent did I think that through technology we could control nature and ultimately "solve" the problem of climate change? Did I *truly* believe in the potential to change the fossil-fuel-driven economic growth model on which our whole way of life was based?

Although my thoughts, and those of countless environmentalists, about the natural world had been strongly influenced by great nature writers such as Thoreau, Muir, Leopold and many others, the exposure to environmental philosophy and psychology opened up a new interior dimension to

> The deep ecologists and eco-psychologists provided new ways to examine the roots of my belief systems

understanding my own motivations. The deep ecologists and eco-psychologists provided me with new ways to examine the roots of my belief systems and my subsequent actions in the world.

I also suspected they held great potential to apply to research about sustainability leadership. For first time, I grasped that how we specifically think about *ourselves* in relationship to nature, and the extent to which we are consciously aware of our interdependence with nature, could determine how we act as corporate executives, entrepreneurs and, ultimately, as sustainability leaders.

Earlier that summer, I had attended a corporate sustainability conference in Chicago entitled "The Sustainable Manufacturing Summit". The conference was one of the first of its kind and was attended primarily by senior sustainability executives from multinational companies, along with a few senior government officials and NGO executives.

The conference was relatively intimate and I had the opportunity to meet senior sustainability executives from companies including SC Johnson, MillerCoors, Motorola, Sprint, InterFace, Mars and Seventh Generation, and a handful of social entrepreneurs pioneering innovative manufacturing approaches to sustainability. Companies such as TerraCycle, the New Jersey-based company developing the innovative waste reduction practice of turning discarded packaging from some of the world's largest consumer products companies into new products.

Based on their corporate sustainability reports and conference presentations, I gathered further evidence about how global companies were implementing sustainability initiatives into their supply chains, procurement, manufacturing, transportation and disposal practices. I methodically analysed how they were reducing their carbon footprints through innovative renewable energy projects using new solar, wind, hydro and geothermal technologies.

I also developed a clearer understanding of many ways they were measuring and reporting on these initiatives. I fleshed out the business cases for sustainability, ranging from energy savings, product innovation, increasing market share and employee engagement.

However, I was looking for something else. I was interested to learn more about the individuals inside these companies who were championing these initiatives behind the scenes. I was curious about their backgrounds, their values and their personal stories. I wanted to know how they became who they were.

The ideas of deep ecology and eco-psychology opened up a new avenue of exploration into sustainability leadership development at the individual level. More specifically, they pointed to a hidden source where the motivation for sustainability leadership came from. I was interested in what they could tell us about motivation for sustainability leadership at the highest levels of multinational corporations.

These two parallel curiosities, the first about the backgrounds and deeper motivations of sustainability executives, and the second about what deep ecology and eco-psychology could tell us about their leadership, came together when I decided to pursue an interdisciplinary social science doctorate in Human Development and Organizational Systems. This in turn led to my focus on worldviews and drove the next five years of my research into how the social sciences could inform our understanding of business leadership for sustainability.

My research framework evolved to include how worldviews are interpreted through eight distinct academic disciplines: deep ecology, eco-psychology, social psychology, developmental psychology, environmental sociology, ecological economics, indigenous cultures and the new field of integral ecology. My thinking was strongly influenced by developmental psychology research about how we actually construct our worldviews, how we interpret the world around us, and how this can change over the course of our lives.

During the same period, I attended more than a dozen corporate sustainability conferences and conducted follow-up phone interviews with 75 global sustainability leaders. I had informal conversations with dozens more. All told, I spoke with more than a hundred sustainability executives over five years.

I also continued to spend time in the forest with natural science and permaculture teachers. The more time I spent with them studying nature, the more I sensed that the sustainable business movement, as it's currently framed and communicated, would never go far enough. Without a deeper understanding of how we human beings

> The more time I spent studying nature, the more I sensed that the sustainable business movement would never go far enough

interpret the ecological systems within which we all exist, sustainability leadership may achieve only to slow down our unsustainable practices.

As sustainability educators, we need to cultivate a new, more holistic curriculum for sustainability leadership in the coming decades. As sustainability executives and activists, we need to comprehend better how an understanding of worldviews can unlock new possibilities for leadership and success in our work. I offer the following pages in the hope they make a contribution to this effort.

Part 2: Exploring the corporate eco-psyche

The major problems in the world are the result of the difference between how nature works and the way people think.

Gregory Bateson

3
Perspectives on ecological worldviews

As midsize mammals dependent on the Earth's ecosystems for life, human beings now face the most serious and complex set of ecological problems in our history. Driven by our ecologically unsustainable way of life and the dramatic increase in our global population, these problems include an increasingly less predictable climate and a wide range of interrelated social, environmental and economic problems.

Although multinational leaders have been immersed with scientific information describing the ecological crisis, overall it has not altered the short-term economic approach to business that is responsible for the serious problems we face. It appears that more information from the natural sciences is

Perhaps the social sciences can now make a vital contribution by reframing ecological issues, especially for sustainability leadership?

Discipline	Theorists	Key concepts and themes
Eco-psychology	Roszak Hillman O'Connor Conn Kahn Doherty	Psyche and Gaia Anthropocentrism of psychology Eco-psychology, health and wellbeing Developing sensory awareness Human relationship with technology Environmental identity
Deep ecology	Naess Sessions Drengson Devall Fox Macy Abram Capra	Deep vs. shallow ecology: ecological maturity Ecological self Technocratic vs. planetary-person paradigm Ecocentricism vs. anthropocentrism Ecological transpersonal philosophy Ecological self: paradigm shift Ecological embeddedness Systems thinking: web of life
Environmental sociology	Dunlap Hedlund-de Witt Bragg Kempton	New ecological paradigm Integral worldview framework Constructionist theory/expanded self-concept Environmental values
Social psychology	Bateson Koltko-Rivera Ray and Anderson	Ecology of mind Hidden nature of worldviews Cultural creatives
Ecological economics	Beddoe et al. Costanza Daly Meadows	Redesign dominant socioeconomic regime Valuing natural capital Steady-state economy, pre-analytic vision Limits to growth
Integral ecology	Esbjörn-Hargens O'Brien Hulme Hedlund-de Witt	Ecological selves framework Reframing climate change debate Social meanings of climate change Integral perspective on worldviews
Indigenous studies	Four Arrows (Jacobs) Hart Pewawardy Cajete	Indigenous education/paradigm shift Indigenous worldviews and social work Indigenous worldviews as eighth intelligence Storehouse of ancient environmental wisdom
Developmental psychology and sustainability	Boiral et al. Brown Rogers Lynam Visser and Crane Rimanoczy Divecha and Brown	Action logics and environmental leadership Conscious leadership for sustainability Corporate ecological selves A geography of sustainability worldviews Sustainability change agent archetypes Embedding sustainability mind-sets Integral action logics for sustainability

TABLE 3.1   Partial list of disciplines, theorists, and themes that make up the ecological worldview literature

not enough. Perhaps the social sciences can now make a vital contribution by reframing ecological issues, especially for sustainability leadership?

There are at least eight interrelated social science disciplines that can be applied for this purpose. These disciplines, which I characterize as ecological worldview traditions, include deep ecology, eco-psychology, environmental sociology, social psychology, ecological economics, indigenous studies, integral ecology and developmental psychology.

Although these schools of thought have existed for several decades as separate academic disciplines, there has not been enough exploration of how they can be integrated into the business curriculum to comprehend the broader implications of sustainability leadership. This is due in large part to the continued silo nature of our universities, with business schools still being separate from departments of environmental studies, psychology, philosophy, sociology, anthropology, communications and the humanities.

As we will see throughout this book, I draw on key concepts and themes from within these disciplines as part of the theoretical framework for my research about sustainability leaders. A partial list of scholars and themes from these traditions is provided in Table 3.1. A brief description of how each discipline has interpreted the worldview concept follows.

> There has not been enough exploration of how these disciplines can be integrated into the business curriculum to comprehend the broader implications of sustainability leadership

3.1 Through the lens of social psychology

As a foundation, let's first explore the concept of worldview, what it means and why it holds so much potential for advancing sustainability leadership. To begin, here are several descriptive definitions and interpretations of worldviews from three social psychologists.

> We continually use our worldviews to make sense of the social landscape and to find ways towards whatever goals we seek

Michael Anthony Hart, social work professor at the University of Manitoba, observes that we continually use our worldviews to make sense of the social landscape and to find ways towards whatever goals we seek. He adds that our worldviews are developed throughout our lifetimes through socialization and social interaction. Hart observes further that our worldviews are encompassing and pervasive, yet they are unconsciously and uncritically taken for granted as the way things are.[19]

Psychologist Mark Koltko-Rivera defines worldviews as a set of assumptions about physical and social reality that can have powerful effects on our cognition and our actions. He describes worldviews as our total outlook on life, society and its institutions. Highlighting the hidden nature of worldviews, Koltko-Rivera adds that explicit references to worldviews are often obscured in academic articles and in our conversations, because they are called other things including values, paradigms and beliefs.[20]

In their seminal book, *The Cultural Creatives*, sociologist Paul H. Ray and psychologist Sherry Ruth Anderson situate worldviews in the context of values and culture in the United States. They characterize our worldview as comprised of everything we believe is real. This includes our understanding and interpretation of the economy, technology, the Earth, how we work and what we value.[21] Ray and Anderson observe that a change in worldview don't happen often because it changes *virtually everything in our consciousness*. They describe this is as a sense of who we are and what we are willing to see, including our priorities for action.

> Our worldview is comprised of everything we believe is real

3.2 Through the lens of developmental psychology

Over the last four decades, researchers from the field of developmental psychology have periodically used the term "worldview" synonymously with "stages of development", "action logics", "orders of consciousness" and "ways of making meaning". Temple University professor Willis Overton characterizes worldviews as narratives that ultimately evolve to become an overarching paradigm constellated by a set of interwoven and coherent set of concepts.[22]

Leadership consultant and human development researcher
Barrett C. Brown puts the worldview concept into a
sustainability context by explaining how developmental
psychology can be used to map how worldviews change
over time and become more complex as an individual's span
of care grows. He observes that as human beings we have
the potential to develop from caring for ourselves/family/
group/nation, to eventually caring for all human beings, and
ultimately to all sentient life. In the context of sustainability
leadership, Brown highlights the importance of effectively
communicating to different worldviews. He explains that in
order to enhance our effectiveness as sustainability leaders we
need be able honour all worldviews as they are, even if they
differ from our own.[23]

3.3 Through the lens of integral ecology

Integral ecologists Sean Esbjörn-Hargens and Michael
Zimmerman put worldviews into a developmental
and ecological context. They observe that within the
environmental movement there are three large groups of
worldviews known as traditional, modern and postmodern.
They explore the implications of worldviews on politics,
religion and culture from a developmental perspective
and explain how environmental issues arise differently for
people depending on their worldview. Esbjörn-Hargens
and Zimmerman observe that this can prevent leaders from
effectively communicating with each other and attribute

the failure of solving the world's major environmental problems at least partially as a failure to differentiate among worldviews.[24]

Now with an initial sense of how the concept of worldview has been described and how it may apply to sustainability leadership, we now focus more closely on the concept of **ecological worldviews**.

3.4 Ecological worldviews

Ecological worldviews can be thought of as deep mental patterns and habitual ways of seeing our relationship to the natural world. Ecological worldviews can also be thought of as the cognitive and perceptual capacity to see the world through the lens of ecology, which is essentially the relationship between species and their environments. This includes the way we think about our individual relationship with nature as well as the relationship between human society, technology and nature.

> Ecological worldviews can be thought of as the cognitive and perceptual capacity to see the world through the lens of ecology

Ecological worldviews were described as early as the 13th century by St Francis of Assisi, who said that all humans were responsible for protecting nature as part of their faith in God. Ecological worldviews have been described in the 19th century through the transcendentalism of Ralph Waldo Emerson and Henry David Thoreau. In

the early 20th century, they can be found in the writings of
the mystic Thomas Merton and land conservationist Aldo
Leopold. In the second half of the 20th century, philosopher
Lynn White explored the larger spiritual implications of
ecological worldviews by attributing their absence as a root
cause for the ecological crisis.[25]

3.4.1 Deep ecology

In the early 1970s, Norwegian philosopher Arne Naess
introduced ecological worldviews into the contemporary
academic dialogue through the philosophy of deep ecology.
Writing from a stone hut high in the mountains of northern
Norway, he articulated a key distinction between shallow and
deep approaches to environmentalism based on the capacity
to access what he called the ecological self.[26] Naess and his
fellow deep ecologists argued that a shallow environmental
worldview of continuous economic growth and controlling
nature through technology must yield to a deeper ecological
worldview where man recognizes his ecological niche among
thousands of species on Earth. Although the philosophy
of deep ecology has gained little traction in mainstream
academia, and essentially been non-existent in the business
curriculum, its call for a large-scale transformation of the
worldviews underlying our short-term economic approach
to business is more relevant than ever in an effort to reframe
sustainability leadership.

3.4.2 Eco-psychology

The connection between human psychology and the health of the natural environment is a foundational assumption of the field of eco-psychology. In this way it offers great relevance for a deeper understanding of ecological sustainability. Theodore Roszak coined the term eco-psychology in 1992 and observed that, in order for human beings to perceive their ecological interdependence, we need to rethink our scientific and anthropocentric worldviews.

> The connection between human psychology and the health of the natural environment is a foundational assumption of the field of eco-psychology

Roszak characterized our collective psychological theories as commitments to understand the interior dimensions of people in various ways. Conversely, he described eco-psychology as a field of inquiry that commits itself to understanding human beings as biological actors on a planetary stage. By observing that all human beings are shaped by the biosphere within which we live, he illuminated how an ecological worldview is part of our potential.[27]

Lewis and Clark professor and eco-therapist Thomas Doherty suggests that the entire field of psychology be reinvented as an ecological discipline and concludes that eco-psychology be considered not as a discipline, but as a social movement and worldview.[28] I agree with Doherty's assessment and would add that it should also form part of a new foundation for sustainability leadership.

3.4.3 Environmental sociology and a new ecological paradigm

More than a half-century ago, the physicist Thomas Kuhn first used the term "paradigm shift" to describe how, throughout history, large-scale crises can serve as a catalyst for transformative scientific change in the world.[29] By acknowledging that nature is a force not to be subdued, but understood and aligned with, Kuhn demonstrated an ecological worldview. Several decades later in the late 1970s, environmental sociologists Riley Dunlap and Kent Van Liere combined the paradigm concept with distinctions between anthropocentric and ecocentric worldviews to conceptualize a **new environmental paradigm**. Influenced by the publication of *Limits to Growth*[30] as well as the philosophy of deep ecology, they described their motivation as stemming from a shift in their own worldviews from the existing dominant social paradigm to a new paradigm based on an ecological worldview.[31]

Dunlap and Van Liere introduced a new survey instrument that was widely adopted in the research community during the 1980s and produced numerous studies. Then, based on new information from the natural sciences and the growing awareness of deforestation, loss of biodiversity, and climate change, they expanded their instrument in 1992 and renamed it **The New Ecological Paradigm Scale (NEP)**.

The idea of a new ecological paradigm has generated significant dialogue over the last three decades. However, findings about ecological worldviews based on the NEP survey have been criticized from a psychological perspective. The major criticism of the NEP is that it is primarily a

measure of values and beliefs on a societal level and does
not get at deeper individual eco-psychological constructs,
as Dutch environmental sociologist and sustainability
researcher Annick Hedlund-de Witt has noted.[32]

By observing that approaches to environmental issues in
academia, public policy and the corporate sphere have not
integrated interior psychological
perspectives, Hedlund-de Witt highlights
the potential for worldview research in
the context of sustainability. Under her
concept of worldview she bundles
together psychological dynamics,
emotional responses and cultural values.
Hedlund-de Witt makes an important distinction between
environmental attitudes and the deeper construct of
worldviews. She concludes that the more encompassing
concept of worldview refers to the foundational assumptions
and perceptions regarding the underlying nature of reality.
Hedlund-de Witt observes that research into worldviews has
historically been underemphasized and calls worldviews a
concept *whose time has come.*

> Research into
> worldviews has
> historically been
> underemphasized

3.4.4 Ecological economics

At around the same time Dunlap and Van Lear were revising
their survey instrument in the early 1990s, University of
Maryland professor Herman Daily was making an attempt
to insert his ideas about ecological economics into the
World Bank's World Development Report. In his 1996 book
Beyond Growth, Daly described a type of ecological or

pre-analytic worldview as part of his theory of steady-state economics.[33] He recounts how he tried to get the simple idea that the economy was a subset of the biosphere into the World Development Report. Unfortunately, his efforts were unsuccessful.

In 1994, Daly left the World Bank to return to academia to advance his ideas about a steady-state economy and a world with limits to growth. He has been writing about ecological economics ever since. Echoing the philosophy German economist E.F. Schumacher as described in his prophetic book *Small is Beautiful: Economics as if People Mattered*, Daly and numerous other ecological economists over the last two decades have conducted significant peer-reviewed research that makes the real cost of depleting our natural capital abundantly clear.

In a recent article about sustainability roadblocks, ecological economists Rachael Beddoe, Robert Costanza and their colleagues offer another perspective on why ecological worldviews are so deeply needed. They describe that underlying our current social and environmental system is a set of interconnected worldviews that support the goal of unlimited economic growth and consumption as a proxy for the quality of life.[34]

In their recent book about ecological economics and limits to growth, Rob Dietz and Dan O'Neill observe how "the pursuit of a bigger economy is

> Our current social and environmental system is a set of interconnected worldviews that support the goal of unlimited economic growth and consumption as a proxy for the quality of life

undermining the life support systems of the planet and failing to make us better off ... The model of more is failing both environmentally and socially ... And practically everyone is still cheering it on."[35] Unfortunately, the many ways that the field of ecological economics can contribute to sustainability leadership have remained largely unexplored in the business curriculum.

3.4.5 Indigenous worldview

The contribution that a psychological, practical and less political examination of indigenous worldviews can make towards understanding ecological sustainability has also been underrepresented in the mainstream business curriculum. In particular, the indigenous principle of an awareness of one's bioregion can form the foundation for an ecological worldview and a strong context for sustainability. By describing how a sense of relatedness and interconnectedness with the natural environment is an integral part of indigenous worldviews, indigenous scholar Four Arrows (aka Don Trent Jacobs) observes that indigenous worldviews align with sustainability priorities and contribute to ecological awareness.[36]

Four Arrows characterizes this sense of relatedness as focusing on how to live in harmony with nature. He states that a sense of one's place is the central theme of indigenous learning and that the very concept of indigenous means "to belong to a place". He describes how indigenous

> The very concept of indigenous means "to belong to a place"

perspectives pay attention to the larger connections rather than maintaining a reductionist view of the world.[37]

By describing the historic capacity for indigenous peoples to listen closely to the land and carefully observe nature, eco-psychologist and phenomenologist David Abram invites us to further appreciate how indigenous culture aligns with an ecological worldview.[38] Abram describes our capacity for observation as the part of us that can feel the wind blowing through the trees and "notice the way a breeze may flutter a single leaf on the whole tree, leaving the other leaves silent and unmoved, or the way the intensity of the sun's heat expresses itself in the precise rhythm of the crickets".[39]

3.5 Ecological worldviews: a missing perspective for sustainability leadership

By reflecting on these various interpretations of ecological worldviews, we can begin to appreciate the potential they hold to advance the field of sustainability leadership and break through existing limitations. This can relate to motivation for, or resistance to, sustainability. A fuller understanding of ecological worldviews also explains a great deal of the conflict society has faced in regards to environmental issues in general, especially in regards to climate change. In a recent article,

> A fuller understanding of ecological worldviews also explains a great deal of social conflict in regards to environmental issues

University of Michigan professor Andrew Hoffman describes how the key to engaging the debate around climate change and sustainability is by addressing our deeper ideological, cultural, and social filters ... In other words our worldviews.[40]

In the minds of corporate executives, ecological worldviews drive deep motivation and form the psychological foundation for careers in sustainability, as we are about to see.

4
Life experiences that shape ecological worldviews

With the preceding overview of ecological worldviews and their implications in mind, I'll begin to share with you what I found from my interviews with sustainability executives. I'll do this directly through their words that I've organized into four major themes throughout the book.

Of the 75 executives that I interviewed, 54 held senior-level positions in multinational companies at either the chief sustainability officer, vice president, director or manager level. There were three CEOs of public companies, six presidents of private companies, six senior executives of environmental NGOs, and six sustainability consultants. A complete description of my research methodology and participants is included in Appendix B.

The first major theme that stood out from my interviews was the majority of executives who described distinct life experiences that shaped their ecological worldview. They often shared stories at length when I asked about their backgrounds and motivations for their work in sustainability. Many of them became animated when telling stories about their childhoods, their families, their teachers and their experiences abroad in developing countries. They traced the origins of their ecological worldviews back to specific points in time, people, places or events that made a significant impression on their lives. Collectively, these stories illustrate that how we think about nature can drive deeper motivation for careers in corporate sustainability.

4.1 Family of origin and early childhood experiences in nature

These first of the five life experiences involved **their family of origin and early childhood experiences in nature.** In response to initial background questions, phrases such as "growing up", "how I was raised", "ever since I can remember" and "when I was a kid" appeared in many of the interview transcripts. These first three interview excerpts all point to how early childhoods in rural environments influenced worldviews. Each of these three mid-career senior sustainability executives attributed their early ecological worldview and eventual career paths in sustainability to their childhoods:

> Growing up my family had a very sustainability-minded approach. My parents were composting and re-using grocery bags before it was

mainstream. My parents were much ahead of their time. I grew up that way. I also spent a lot of time outdoors and developed a deep appreciation for nature … So I think it influenced me to become an environmentalist.

When I was a kid my grandfather had an apple orchard. We would spend summers going from one grandparent to the other. We just played outside in the apple orchard that was maybe 60 acres or something. There were all these cows around. You just kind of learned about the role of growing food in a way that just kind of enveloped me … That was just how I was raised.

I was raised in a rural, small town in Vermont and was like most boys in rural America at the time. I was outside all the time. I also did a lot of fishing and hunting when I was young. My father's family influenced my upbringing. We gardened a ton. I did canoe trips with scouts and all that sort of thing.

In addition to going to the park with his father at the age of six, this president of an environmental NGO shared how the careers of his two parents and his older brother influenced his thoughts about the environment while growing up in California during the 1960s:

I think it started when I was about six years old and my Dad would take me to the park. I grew up around Modesto, California. My mom was in democratic politics and my dad was a teacher.

> My older brother made a film in the early '70s
> about garbage. I watched all the development in
> California and I knew that it did not feel right.

An eco-psychological perspective can be found in this next quotation:

> Ever since I can remember, I've been happiest
> when I'm out in the wilderness. I grew up walking
> along trails in the foothills near my house ...
> One day I started walking fairly slowly. And I
> just started seeing so much more. Going slow,
> listening.

As a final example of this first finding, here is a reflection about childhood that also leads into the second finding about environmental education. It is from an interview with the president of a national consumer food company. In it he describes an unforgettable experience in middle school:

> I grew up in West Nyack, New York along the
> banks of the Hudson River, less than an hour
> north of Manhattan. The teachers at the middle
> school I went to had a mission to get more
> environmental awareness into the classroom. So
> they taught us about why the Hudson was so dirty,
> told us about all the industrial dumping from
> factories up the river. Then one day they took my
> class out on the *Clearwater*, Pete Seeger's sailing
> vessel that was dedicated to cleaning up the river.
> I remember how they used nets to dredge up and
> remove garbage from the river. I saw stuff like old
> tires, pieces of cars, old luggage and lots of scary

stuff. We also sang all the Pete Seeger and Woody
Guthrie songs like *This Land Is Your Land* and
Inch by Inch and somehow through the music
and seeing all the junk in the river, it made a big
impression on me.

4.2 Environmental education and memorable teachers and mentors

The second significant life experience that sustainability
executives brought up when asked about their motivation
for sustainability was specific parts of their environmental
education and the memorable teachers and mentors they
studied with during this time. Many of the executives
reflected on experiences in college or graduate school. For
instance, this executive, a vice president at a global consumer
food company, remembered one particular class he took that
was based on the systems thinking work of Thomas Lovejoy,
a widely respected tropical biologist:

> I went to Kenyon College. I remember that I had
> to take at least one natural science class. There
> was a class on systems thinking based on Thomas
> Lovejoy's work and the value on standing forests.

The president of a consumer products company spoke at
great length about his discovery of eco-psychology while an
undergraduate:

> I was a psychology major at Stanford. My senior
> year I heard about the field of eco-psychology.

There wasn't anyone who taught it at Stanford but I found a professor at UC Berkeley who had edited an anthology. So I got together with him and did an independent study. I lived in a co-operative on campus called Synergy. They had a big garden outside the house and that is when I first became really interested in how to build soil and grow food … My perspective was being shaped by the time I was spending with the farmers and became the impetus for my work in sustainability.

This executive with a global NGO described how his senior thesis opened up what he referred to as his theme of the integrated nature of disciplines:

I went to Brown and studied environmental science. Part of the curriculum was to write a senior thesis. My thesis explored what it would be like if ranchers were ranching native animals instead of cattle on western rangelands. I explored what that would look like. For me it was the beginning of blending ecological science with culture and economy. Since then it has been a real exploration into that blending of disciplines, which for me has been a key theme since I went to college, the integrated nature of disciplines.

This executive refers to a philosophical shift that took place during college and how social activism inspired his future work as a sustainability leader.

Coming out of high school, I was not the person I now am. I trace back to college as a turning

point. Soon thereafter I became a different person,
I would say. I grew up in a fairly sheltered upper
middle-class suburb of New York ... There was
something about breaking through the bubble
arriving to college ... that everything was fresh
and new ... There was a lot more to think about
and read about in college and then there were
also ways to apply idealism through community
service, which was a big piece of it ... Then the
pivot from service to activism was one I was
feeling out as I was later going through college,
junior and senior year. I had a couple of professors
who had been activists themselves in the '60s.
Stories of student turmoil and protest movements
really got my attention and interest.

4.3 Seeing poverty and environmental degradation in developing countries

The frequency with which sustainability executives reported
the impact of seeing poverty and environmental degradation
in developing countries was surprising. When I asked them
where their motivation came from or what influenced their
worldview, many of the executives shared stories about
their experiences in developing countries. Several executives
worked in the Peace Corps or other volunteer organizations
in South America and Africa early in their careers. They
reported how seeing poverty and environmental degradation
at first hand had a significant impact on their worldview.

For instance, this participant described how a volunteer experience in South America changed his life:

> I went to Paraguay in the summer of 1991 in-between my junior and senior year in high school ... I lived with a family in a very rural part of the country ... Every few kilometres there were tiny shacks where families lived beside their fields. Mostly they were growing single crops like soy and cotton. There were big open fields for cattle created by clear-cutting. In the distance you could see a stand of old-growth forest but it felt like it was always in the distance. The deforestation was depressing. I remember feeling a lot of sadness about what I saw.

The next comment is from a long-time executive with a large coffee company. As part of their company's policy, employees are selected to attend immersion trips to the countries of origin where the coffee is grown. Here is how he described his experience:

> When I first went to Costa Rica in 1992 I did not see any poverty. Then I took a week's vacation in 1995 and travelled to northern Guatemala and southern Mexico and saw all the poverty. I used my own vacation time and paid for the trips myself. I lived with families, took a total immersion language course, and became more and more passionate about these issues. I came to understand the struggles and became so committed that I did a lot of this on my own time.

Along similar lines, this executive spoke about how his travel and work in Central and South America allowed him to formulate new thoughts about sustainable development, social justice and the environment.

> I was able to get to the developing world early in college through an internship. I think this is where my interest about poverty and inequality issues in the United States pivoted to become more global … I became aware of how environmental and social justice issues went in tandem. Then it prompted travels in Bolivia for my senior thesis research and later living and working in Nicaragua for half a year right after college … I learned more about the questions that I needed to be asking more than getting answers … realizing that people and the environment are very much intertwined.

4.4 Perceiving capitalism as a vehicle for environmental or social activism

There were a number of executives that described moments during their careers when they first perceived capitalism as a vehicle for environmental or social activism. It was surprising how many senior sustainability executives at multinational companies had extensive prior experience working in either environmental NGOs, the public sector or both.

These executives described similar versions of stories where they had started their careers motivated to work on a combination of social justice or environmental issues. Then, after a number of years, they intentionally decided to move into the private sector as a way to leverage their experience and have what they perceived to be a bigger impact on the world.

For example, this participant, a widely respected senior sustainability executive who has worked for two multinational corporations and pioneered many corporate sustainability practices, described how he went between the public and private sectors earlier in his career:

> I started my career working for Bernie Sanders on national budget and defence issues. Then I went to work for Ben Cohen and Jerry Greenfield. We were able to figure out how we could take a peace dividend ... Later I went on to Greenpeace with a clear focus on global warming ... I ended up deepening my understanding of corporations and developing a new model of corporations as a more positive force in the world ... I came to the conclusion that being part of a corporation was how I could have the biggest impact.

This senior sustainability executive, at a multinational consumer products company with one of the largest global supply chains, described how he spent more than two decades in the public sector before moving into his role in corporate sustainability:

> I went to college in Colorado and was involved in the protest to shut down the Rocky Flats nuclear power plant. I went to work for Senator Tim Wirth and wanted to help end the nuclear

arms race … I remember Tim saying not on his
life was he going to compromise. He said he
was going to fall on his sword before he ever let
nuclear weapons continue. When the Berlin Wall
fell, Tim kind of pivoted from the east to west
political military issues to the north to south
environmental and social issues and I pivoted with
him. It was a turning point for me in terms of
commitment to the environment.

4.5 A sense of spirituality and service

The fifth significant experience that sustainability executives
described as part their motivation for sustainability involved
a sense of spirituality and service. For example, this
senior sustainability executive at a global communications
company told a story about how she grew up with nature in
her backyard:

I believe this whole area of environmental
corporate activism also involves spiritual
development. I grew up on a creek in Sioux City,
Iowa and just that experience gave me a love of
nature. Ever since I was a child I wanted to serve
and give back to the community. At this point in
my life, I can't imagine having a more satisfying
career because my spiritual aspect is being
addressed through my work in sustainability.

This next executive, at a global wood products manufacturing company, shared this very personal reflection of his spirituality during his interview:

> I am very much of the view that we are all parts of a very interconnected, interdependent whole. All species, or natural features as Joanna Macy puts it, are all important; we all have our place; we are all worthy of respect. However, humans have set themselves apart and above nature to everyone's detriment. Technology has only increased this divide ... My spirituality is nature-based ... it is definitely tied with respect, awe and gratitude for nature. In nature is where I am more at home. I was fortunate to have had access to nature in my childhood. I've always had a deep connection and humility. My work in sustainability has only enhanced and deepened my perspective.

Along the same lines, this executive at a national waste management firm shared how his experience with transformational shadow work helped deepen his sense of the connection between spirituality and sustainability:

> I was the first executive director of the Mankind Project, which gave me a lot of perspective on deep transformational shadow work. I suppose that I've been working at the intersection of spiritual development and sustainable business practice ever since. At this stage of my career, spirituality, sustainability and work are interwoven.

This executive at a global footwear and apparel company shared her awakening environmental consciousness in the following way:

> I read *The Ecology of Commerce*[41] and listened to Paul Hawken speak. I also came across the Natural Step. It became apparent to me that we were operating against nature's rules.

This last quotation from an executive at a major software company cited the movie *An Inconvenient Truth* as a pivotal moment that catalysed and deepened her commitment to sustainability:

> That would be when I saw *Inconvenient Truth*. I was getting increasingly alarmed at what was and wasn't happening in the areas addressing climate change and thought that I needed to personally get more involved. I became involved with an environmental group in my community and this kind of grass-roots effort brought more visibility to what you could do personally and what we could do collectively ... so I started looking for a job in the company to address that.

4.6 Initial reflections

Although my research was not designed to uniformly examine the biography or lifeline for each of the participants, a chronological and possible developmental sequence did suggest itself within the five significant life experiences. The first set of quotations about how family and early childhood had a significant impact on their worldview generally referred to the K-12 years.

The second set of quotations that focused on environmental education generally corresponded to their college and graduate school years. This appears to have been a period of great discovery in their lives. However, what distinguishes these sustainability leaders from many other people is that this period appears to have contributed significantly to the formation of an **ecological worldview.** Many of them described a time when they first cultivated an understanding of ecological systems and social justice. Many of them then began to orient themselves towards eventual careers in sustainability.

The third finding about experiences while travelling in developing countries also appears to have been a period of great discovery. These experiences are closely related to the second finding, although they appear to have taken place in many instances a bit later in their lives. Many participants described how witnessing extreme poverty and environmental degradation for the first time while in developing countries appears to have deepened their ecological worldview and commitment to their work in sustainability. In some cases these experiences appear to have been epiphanic in nature.

The fourth set of quotations that relates to career shifts appears to represent an evolution of their ecological worldview. These experiences often took place after a decade or more in careers in the public sector or at an NGO. It reveals how many of the participants became aware of the impact of global corporations on planetary ecosystems and global social issues. While most likely personal financial gain played some part in their decision, in each case they appear to have made conscious decisions to move to the private sector in order to have a bigger impact on the world.

The fifth significant life experience that sustainability executives described as part of their worldview and their motivation was spirituality and a sense of service. These expressions generally appear to be a reflection of their current lives, which would correspond to their late thirties or forties for the large majority of the executives I interviewed. By evoking a spiritual source and a belief in the intrinsic value of nature as part of their motivation for their work in sustainability, these executives were expressing ecocentric worldviews and a sense of their ecological self, which we turn to in the next chapters.

5
Anthropocentric blindness

Among many themes in the ecological worldview literature, there are two that I found to have great potential to provide new insights into the psychology of sustainability leadership. These are the distinction between **anthropocentric and ecocentric worldviews** and the **ecological self**. These two themes can also be considered as psychological constructs and as components of an ecological worldview. The far-reaching implications of anthropocentric worldviews have received particular attention from the fields of deep ecology, eco-psychology and indigenous studies.

5.1 The pervasiveness of anthropocentric worldviews

Over the last 50 years, social science researchers from numerous disciplines have characterized the worldviews of most people as being anthropocentric, which reflects a belief that human beings can ultimately control nature through technological and economic advances. An anthropocentric worldview is based on a belief that human beings are at the centre of the universe and the most significant species on Earth. It assumes that all phenomena in the world should be interpreted in terms of human values and experiences. A person with an anthropocentric worldview has a more instrumental view of nature.

> The belief that human beings can ultimately control nature through technological and economic advances

The depth of our anthropocentric worldviews is expressed by Swiss archetypal psychologist James Hillman, who lamented that the entire field of Western psychology itself has become so lost on the inner perspectives of human beings that it seems to have forgotten that we are embedded in the natural world.[42] In his 1967 article entitled "The historical roots of our ecologic crisis",[43] the philosopher Lynn White suggested that the anthropocentrism of Western culture and religion lies at the roots of our destructive tendencies towards nature. Today, his observations seem more prophetic than ever.

Nature Conservancy CEO and author of *Nature's Fortune* Mark Tercek offers another way of thinking about anthropocentric worldviews by observing that people often speak about nature, or Mother Nature, but for the most part think of Nature as something separate and apart from

> The green movement is not a movement to save nature, but actually a movement to save the human race

nature.[44] The Canadian novelist and environmental activist Margaret Atwood makes the difference between these two worldviews clear. She observes that the green movement is not a movement to save nature, but actually a movement to save the human race. She adds that nature, in all of its evolutionary biodiversity, will prevail long after the human race, and that humans, an unusual midsize mammalian species, may disappear from the planet.[45]

As pertains to our work as sustainability leaders, anthropocentric worldviews can act as blinders that underlie entrenched resistance to the environmental initiatives we champion. Similar to my realization about the fragility of the Earth's ecosystems, being aware of the extent of anthropocentrism allowed me to see the limits of the current sustainability movement. I became aware how this way of thinking has affected powerful institutions in Western culture since the Renaissance and the so-called "scientific revolution", with its overly mechanistic interpretation of reality.

Most unfortunately, anthropocentrism has been a severe limitation of most ways we approach technology and innovation. Perhaps the most glaring example is the recent unprecedented expansion of hydraulic fracturing technology, or "fracking", in the United States during the 21st century.

5.2 Ecocentric worldviews

Conversely, an **ecocentric worldview** expresses a belief that human beings are dependent on, and literally embedded in, the Earth's ecosystem. An ecocentric thinker sees the Earth's biosphere at the centre, with *Homo sapiens* as one of many thousands of species that are dependent on the Earth's living systems for survival. An ecocentric worldview means having a basic understanding of non-human organisms and the Earth's ecosystems. A person with an ecocentric worldview maintains a more intrinsic and spiritual view of nature.

> An ecocentric thinker sees the Earth's biosphere at the centre, with *Homo sapiens* as one of many thousands of species that are dependent on the Earth's living systems for survival

Psychologist Daniel Goleman, author of *Ecological Intelligence*[46] and *Eco-Literate*,[47] observes that an ecocentric worldview requires that we apply what we learn about how human activity impinges on ecosystems so as to do less harm and live sustainably in our ecological niche. Francis Moore Lappe, in her powerful treatise *EcoMind*, points towards the ultimate power of ecocentric worldviews by explaining, "as we rethink the premises underlying this worldview, we move to a different place altogether—a place where we experience ourselves and our species embedded in nature ... With an eco-mind, we move from 'fixing something' outside ourselves to re-aligning our relationships within our ecological home."[48]

By far the oldest expression of ecocentric worldviews comes from the many indigenous cultures around the world. Boston psychotherapist Leslie Gray notes that indigenous worldviews can be traced back continuously for as much as 40,000 years. Considering that the modern disciplines of economics, psychology, and sociology are all less than 300 years old, we can appreciate the relative longevity of indigenous worldviews. Inspired by the ecofeminism of Joanna Macy, Gray points out that there are many models of sustainable indigenous societies, whereas there are essentially no models of sustainable industrial societies. She laments the tragedy that we are wasting all this accumulated knowledge from the more than 300 million indigenous people living in the world today.[49]

In *Unlearning the Language of Conquest*, Rutgers University philosophy professor Bruce Wilshire states that one of the primary contributions of an indigenous worldview is that it may allow us to remember ancient human knowledge and intuition that emerged directly from observing nature, wisdom that has since been compromised by too much analysis and cognitive thinking in our modern industrial and technological world. In essence, he is saying that our capacity to closely observe nature has also been severely compromised by the anthropocentric roots of our modern disciplines of economics, psychology and business.[50]

> An indigenous worldview may allow us to remember ancient human knowledge and intuition that emerged directly from observing Nature

A third perspective on the relevance of indigenous worldviews is offered by Professor Greg Cajete from the University of New Mexico. He characterizes indigenous worldviews as a storehouse of ancient environmental wisdom and observes that we've reached a time in human evolution where we need to return to a **unified path of relationships**.[51]

As sustainability leaders, it is vitally important that we understand the pervasiveness of anthropocentric worldviews and work on new ways to overcome this social and psychological phenomenon. As part of a new psychology for sustainability, the capacity to cultivate ecocentric worldviews may hold the potential to break through many of the most entrenched psychological barriers to change.

> The capacity to cultivate ecocentric worldviews may hold the potential to break through many of the most entrenched psychological barriers to change

6
The ecological self

When I was growing up in the hardwood forest of northern New Jersey, I spent most of my free time outside in the woods. My earliest memories are turning over rocks to find salamanders in the glen down the hill from my house. There were box turtles, butterflies and bullfrogs everywhere on the moss-covered rocks. On summer nights the air was filled with fireflies.

Outside my back door there were many big trees to climb. My favourite was a massive red oak, where I spent countless hours sitting on a wide branch high above in the tree. Up in the canopy, nobody could see me. It always surprised me when my mother would blindly call out from the front door up into the tree when it was time for dinner. She would laugh and call me her little chimpanzee.

This is the place and time when I first recognized my **ecological self**, which is the part of us that identifies our

> The ecological self is the part of us that identifies our self as literally part of nature

self as literally part of nature. In this chapter, we explore how the development of ecological self can help us more clearly perceive the Earth's ecosystems within which we are embedded. As sustainability change agents, this can help us communicate in new ways and maintain our resilience and motivation. By creating ways for others to experience their ecological self, we may find new, more effective ways to overcome resistance and access decisive bravery, as my daughter Maggie likes to say.

In an effort to apply the ecological self to sustainability leadership, I focus my attention on the deep ecologists, eco-psychologists and integral ecologists that have written about the ecological self explicitly from a developmental perspective.

6.1 Deep ecology

Norwegian philosopher Arne Naess made the first reference to the ecological self in 1972 as part of the philosophy of deep ecology. Building on his concept of many-sided maturity, Naess observed that a person could be mature in social relations but have an adolescent ecological self.[52] This observation is an important one. It suggests that we are all capable of different stages of awareness in distinct areas of our development, which is at the heart of an integral perspective.

> We are all capable of different stages of awareness in distinct areas of our development

Also during the early 1970s, pioneering human ecologist Paul Shepard contributed to our understanding of ecological self. He captures its essence by describing a potential state of consciousness where the epidermis of the skin is like the surface of a pond with a felt sense that nature is continuous within us.[53] While the capacity to comprehend our ecological embeddedness may be beyond our day-to-day mental habits, it is within our potential should we choose.

Humboldt State sociologist Bill Devall, in the classic 1985 book that he co-wrote with George Sessions, *Deep Ecology: Living as if Nature Mattered*, observed that we underestimate our self-potential by not appreciating our ecological self.[54] Devall contended that the ecological self is part of the transforming process that is required to heal ourselves in the world.[55]

6.2 Eco-psychology

Eco-psychologist Laura Sewall puts forth the idea that the ecological self matures through the recovery and development of our sensory systems, which she calls exquisitely evolved channels for translating the "in here" and the "out there". She recommends five perceptual practices for perceiving our ecological conditions and describes how it is through these practices that our inner and outer worlds can become less rigid and the mature ecological self perceives its permeability.[56]

This involves having a direct experience of the interconnectedness of nature. Ultimately, our empathy for and identity with the broader ecosystem occurs as a result of these changes in perception. Once again, we can see how our perception of the ecological challenges we face as sustainability leaders can be enhanced through the development of our ecological selves.

Australian environmental psychologist Elizabeth Bragg has explored the concept of the ecological self through the lens of developmental psychology. She surmises that an expanded self-concept through the ecological self can affect the functioning of an individual in the environment and explores how self-constructs can be changed. In regards to how an expanded self-concept through the ecological self can affect our behaviour, Bragg points towards the basis for my theory that the development of ecological self can affect our capacity as change agents for sustainability.[57]

6.3 Integral ecology

In Chapter 3, I described how integral ecologists Sean Esbjörn-Hargens and Michael Zimmerman have conceived a theoretical framework of ecological selves based on our capacity to experience multiple perspectives and identify with increasingly complex levels of the natural world.

Their model is presented in the ground-breaking volume *Integral Ecology: Uniting Multiple Perspectives on the Natural World*. In it they provide detailed descriptions of the ecological selves that individuals can hold. By doing so, these two authors have created one of the only typologies of ecological selves and associated worldviews. Their model outlines identifiable patterns of how individuals interpret the natural world and how each pattern affects what an individual can be aware of, reflect on and act on.[58] It is through the connection between an awareness of ecological self and action towards sustainability initiatives that the discipline of integral ecology holds new possibilities to unlock new energy for sustainability leadership.

> It is through the connection between an awareness of ecological self and action towards sustainability initiatives that the discipline of integral ecology holds new possibilities to unlock new energy for sustainability leadership

6.4 Research about the ecological self in the corporate world

From within the sustainability leadership literature, and most closely related to my research, sustainability researcher Katrina Rogers used the ecological selves framework as a theoretical lens to explore how this influenced the worldviews of executives in a single European company and their ability to confront global environmental challenges. Rogers found that executives were able to identify specific moments that led to expanded ways of thinking about the environment.

She observes that while certain executives characterize these changes as epiphanies, others described a more

gradual evolutionary shift. All the executives reflected on
these shifts as being a permanent change in the way they
thought about and approached their work. Rogers reported
that those executives that experienced behavioural changes
demonstrated worldviews that appear to be at the more
advanced end of the ecological selves spectrum.

She found that these executives demonstrate a more
highly developed sense of complexity, systems thinking and
interdependence. Rogers speculates that further use of the
ecological selves framework, and possibly the development
of a new instrument, could lead to new insights and a
deeper understanding about how leaders develop advanced
capacities to understand more fully the role of their
companies within the ecological crisis.[59]

Despite the potential for further insights on the
development of the ecological self to be applied to the field
of sustainability leadership in both academia and in the
corporate world, there has been very little exploration of this
application. To the best of my knowledge, other than the
small sample study conducted by Rogers,
there has been no large-scale empirical
study of corporate sustainability leaders
based on the construct of the ecological
self. University of Washington professor
Peter Kahn and his colleagues have
conducted extensive research on the
human relationship with nature and
technology, but focused primarily on
childhood development and education.[60]
Award-winning journalist Richard

> I speculate that
> most corporate
> executives and
> business educators
> not directly involved
> with sustainability
> have not felt a
> sense of their
> ecological selves

Louv has produced an immense body of work about our relationship with nature and its broader implications, but this is not specifically focused on corporate sustainability leadership.[61]

Although the transformational workshop created and facilitated by Joanna Macy, John Seed, Molly Brown and many others to cultivate the ecological self called **The Council of All Beings** and the **Work That Reconnects**[62] has been spread widely since *Thinking Like a Mountain: Towards a Council of All Beings*[63] was published in 1988, these workshops have not received enough attention within the corporate world and in mainstream business leadership education. I speculate that most corporate executives and business educators not directly involved with sustainability have not felt a sense of their ecological selves.

As facilitators, Joanna Macy and her many skilled colleagues have spent much of their lives' work helping people *feel* and experience their ecological selves. Macy describes how the interdependence of all life remains just a mental concept, without power to affect our action in the world, unless it takes on some emotional reality.[64] By doing so, she makes the important distinction that in order for the ecological self to catalyse a new psychology, and ultimately a new way of acting and making decisions, a future sustainability leader will need to feel it.

> In order for the ecological self to catalyse a new psychology, a future sustainability leader will need to feel it

7
Expressions of ecocentricism and ecological self in the corporate world

By using the two key themes I describe in the previous chapter as deductive lenses for further analysis of my interviews, I uncovered five distinct ways that sustainability leaders express ecocentric worldviews and a sense of ecological self.

Phrases such as "ecological context within which we live", "learn from natural systems", "inherent value in nature", "interconnectedness of humanity and the natural world" and "truly seeing other species" are just some of the examples that appeared during my interviews that are indicative of an ecocentric worldview and the ecological self.

7.1 An awareness of ecological embeddedness

In response to my questions about how they think about their relationship with nature, many executives demonstrated an awareness of the embeddedness of human beings within the Earth's biosphere, one of the key characteristics of an ecocentric worldview and ecological self.

For example, this long-time sustainability executive at a major global apparel and footwear manufacturer described her worldview this way:

> I've always understood at a fundamental level that the economy and society are within the context of the environment. So we really can't do anything without paying attention to the ecological context within which we live.

This next executive, the senior sustainability executive at a company that produces natural household cleaning products, reflected on the potential for biomimicry and industrial ecology to make the world better:

> My awareness of just how much we can learn from natural systems has evolved over time. I continue to look more closely at how biomimicry and industrial ecology could reframe our industrial world and make it so much better. However, I'm not so nature-centric that I don't think that there's a vital role for humans within all this. We possess the ability to control our processes and make them more efficient. However, we have to sit within the natural system and learn from it.

In reference to one of the key principles of natural capitalism, this president of a manufacturing company spoke of "being of service" and "restoring ecosystems". He articulated a specific point in time where he expanded beyond his thinking of himself as just an organizational leader to wanting a better understanding of ecosystems in this manner:

> It was there that I realized I made a shift from being primarily interested in my own experience of being a leader and interpreter to actually understanding ecosystems better in order to be of service and in some way conserve or restore ecosystems.

7.2 An awareness of the vulnerability of planetary ecosystems

During our interviews, many executives demonstrated a heightened awareness of the current vulnerability of the Earth's ecosystems, also indicative of an ecocentric worldview.

This executive, the head of natural resources management at a major global food manufacturer, focused on how she sees her role as a sustainability change agent expanding beyond her own company:

> We are at risk to losing an enormous amount of topsoil and people do not understand that. I am very concerned about water allocation, very

concerned with mono crops. In Oregon, GMO
sugar beets are being grown right next to organic
… This is my passion and I am fortunate that the
company allows me to look at agriculture.

Along the same lines, this executive reflected on his hopes
and concerns for the future by highlighting the health of the
oceans and carbon emissions during our interview:

I hope that the next stage is a broader
understanding of social and environmental equity
as the cornerstone … I have had a bit of a shift in
my thinking. There are so many reasons to limit
the amount of carbon into the atmosphere. The
health of the oceans is a major one. They are
taking a big beating due to acidification to the
point of dying.

This executive, the director of sustainability and
stewardship in a major global technology company, offered
this reflection that suggests a belief of a new ecological
paradigm:

We are at the end of the age of oil. I have
convinced myself by doing the math that if
everyone lived like we do, we would need seven
Earths. So the big question is how are we going
to go from a consumption society to a balanced
system?

While describing the evolution of the priorities of his
company's corporate social responsibility policy, this

executive spoke about the importance of sustainably sourced products to meet the needs of 9 billion people:

> We started out giving money to save endangered species but we now know that we need to influence land policy and how products are sourced. This is now the most important part of our work. As we near 9 billion people we need to meet their needs with less land and sustainably sourced products.

Finally, this executive reflected on the long-term sustainability goal of reducing overall consumption in a steady-state economy, a key concept of both a new ecological paradigm and an ecocentric worldview:

> The longer-term issue that I see in terms of leadership in commercial enterprises is that eventually we have to deal with consumption with a big C, not just Cradle to Cradle thinking, but we have to transition to business models that do not depend on growth, that in fact thrive on the basis of reducing consumption. Right now the business world is getting on board with being more efficient because they know we are running out of natural resources or they are afraid of climate change or political instability, they are on board with this first phase, but not the second phase.

7.3 A belief in the intrinsic value of nature

One of the key distinctions between anthropocentric and
ecocentric worldviews is whether one believes nature is
to serve man or whether it has intrinsic value. This next
participant reflected on this core philosophical question.
Drawing on his background as a senior executive with a
global environmental NGO, he described how the two sides
of this debate are affecting his thinking:

> There's a fascinating debate going on in the
> scientific circles right now. On one side is the
> value of the natural world to human beings that
> reduces it to economic value and human life, and
> risk reduction value. I think what's happening is
> that most people know that this is intuitively a
> narrow view, but it's one that will speak to the
> people that we see at Davos and elsewhere ...
> On the other side is that we not only depend on
> nature, but there is an inherent value in nature ...
> This is the camp of the spiritual values or you can
> say even the intrinsic values of the natural world
> regardless of how it serves human beings. But I
> think where the conflict arises is that in an effort
> to speak to the mainstream, the language is being
> reduced to a story of nature that serves humanity
> through economic and human wellbeing ...
> However, for many of us who have a broader view
> of the interrelationship and interconnectedness
> of humanity and the natural world that is
> problematic.

This next participant referenced one of the iconic environmental books of the 20th century: *A Sand County Almanac* (1949) by Aldo Leopold. By describing the influence of great nature writing on the development of his ecological worldview, he illuminates a core realization of an ecocentric worldview:

> It comes from Aldo Leopold that we need to quit being the lord and master of the world and become a plain citizen of it. We need to truly get away from a human-centric to a more nature-centric, shall we say, view. I don't think we can completely figure out how complex life is. But I do think it is possible to relate to and connect to it. I think it's truly seeing other species at least on an equal plane with us.

Along the same lines, another executive described how reading the nature writers of the 20th century had informed her environmental ethic by adding:

> So when I think about where my worldview comes from I can relate a lot to Aldo Leopold and historical lovers of nature like John Muir.

This next quotation from a long-time sustainability consultant offers yet another vivid expression of an ecocentric worldview and ecological self:

> I'm convinced that humans are an integral part of nature, not masters of or separate from nature, and that through our self-reflective capacities as

human beings we can harmonize our actions with
the natural movements of nature.

7.4 Enhanced systems consciousness

Another distinctive characteristic of an ecocentric worldview
is the capacity to see themselves and their organizations
within the complexity of ecosystems.

For example, this CEO of a corporate environmental
NGO based in Washington, DC, said:

> My personal view is that we've got to find a way
> to move from the goal of just understanding the
> natural environment to the realization that we
> ourselves are causing the environment to change
> drastically around us for the first time in the
> history of man. I think that changes the game. I
> think that I struggle with this personally. I don't
> believe that it is possible at this point to mitigate
> the damage we have started. Now we need to
> mitigate how bad it's going to be, how we're going
> to help people deal with these environmental
> realities in the future.

Another executive, at a national environmental coalition
that focuses on working with members of Congress for
progressive climate and energy policy, shared her perspective
this way:

> Environmental movements take a long time. We
> should not be surprised. What we see is a scaling

up with more sectors, more brands. We were
prepared for companies to back out, but we are
actually getting more calls. The US Chamber
of Commerce and big oil and big coal are
disproportionately influential. If we really look
at history, this is their last gasp. We're seeing a
crescendo of activity that will ultimately result in
a long-term careful solution.

This senior sustainability executive at a major apparel
and footwear manufacturer offered this reflection about
her ecological worldview that demonstrates her systems
consciousness:

I think probably that where I come from in terms
of my ecological worldview is systems thinking
and the interconnection of so much of what we
do and our impact on the environment. I've spent
a lot of time over the years around sustainability
and been exposed to a lot of what's going on in
the world.

When asked how he thought about the impact of his work,
here is how another senior sustainability executive put it:

The next circle out there is the whole planet ...
Quite often it breaks down to understanding
yourself and your dependence on nature. There's
an interrelationship obviously. It means taking
yourself and your team out into the world and
becoming aware of how you are impacting the
bigger ecosystems and making linkages. You could
scale that down or up however you want to. But

it's basically how the social and economic systems
of the human community are in relationship with
ecosystems. You need to have an understanding
of all that kind of stuff in order to be skilfully
engaged.

This participant, the CSO at a global travel services
company, described how the concept of waste had led her to
a deeper appreciation of systems thinking:

When you step into a role like this what you
think will inspire you changes. For example, I
never thought I would be so excited about trash.
However, I realized that I was getting excited
about systems thinking. In order to be a real
change agent you have to understand the whole
system. One day I put on my gloves and went
through the trash in one of our buildings. When I
thought about waste diversion, I began seeing the
entire global waste system.

Another executive, when asked what some of his key
takeaways were since he began his journey as a sustainability
leader, replied:

First, that the more you work on sustainability
you realize it is not just connected to other issues,
but the same as other issues, like ethics, religion,
business, family, education, health, poverty,
respect, government.

This sustainability executive at a multinational food
company described how she and her team see their company

as part of the global food system. She described how they are using their company as a platform to influence other multinational companies towards more sustainable agricultural practices:

> We have a series of position papers that we worked on about a broad range of topics ... From climate disruption to local food systems and the value in terms of big world/small planet point of view ... the finished papers will be external. They are very good papers for conversations. We are using the papers to influence other multinational companies who may not understand what we are doing ... The papers are country specific. The French are very progressive. No genetically modified food. In the next several years we can be a catalyst for safe food ... If we capitalize on our ability to tell our stories and to really engage our consumers, that is where our strength comes.

7.5 Earth-centric circles of identity and care

The capacity to identify with a widening circle of human communities and eventually all species is another important characteristic of an ecocentric worldview.

There were numerous instances where those interviewed indicated a heightened awareness of the entire global community. For instance, when reflecting on the issue of climate change, this chief sustainability officer highlighted in the following way a perspective missing from the political climate change debate:

> Of course from a global perspective climate change is an enormous issue that we should be addressing. But I think one thing that is a little bit absent from these conversations is the outsourcing of our industrial processes to these other countries and our being ignorant of the effect of this. Also the reality that there is so much of just basic environmental protection that is not happening in developing countries. I think that's under-reported. Most people know there's pollution in China and in India, but we haven't included that within our global environmental goals as well as I think it could be.

This director of sustainability spoke of climate change being an issue of equity for people in underdeveloped countries throughout the world:

> Ultimately climate is an issue of ethics and equity, and solving it seems like an obligation to our kids but also to poorer people around the world.

This quotation from a long-time senior sustainability executive at a major apparel manufacturer reflected that our language and culture are still embedded in our patriarchal society.

> I think we are honing our approach. It's an ever-widening circle of learning. The work we're doing on diversity and culture, recognizing what are

the patterns and the artefacts in the culture, I
continue to find it so helpful … How do we lead
going forward? We need to move towards a more
matriarchal society from the dominant patriarchal
societies. We continue to need to see that
sustainability is embedded in the patriarchal and
explore how we can move into a more feminine
nontraditional approach.

7.6 Initial conclusions

The interview quotations presented above offer a more
holistic view of the sustainability leader than what is
currently understood. One can see in these expressions
representations of both their ecological intelligence/cognitive
development, as well as awareness of ecological self/
emotional development. Both form part of their ecological
worldview and the motivation for their work in sustainability.

Phrases used such as "running out of natural resources",
"industrial ecology", "learning from natural systems" and
"biomimicry" all point towards the scientific aspect of their
ecocentric worldview. Whereas, phrases such as "ecological
context within which we live", "inherent value in nature",
"truly seeing other species" and "interconnectedness
of humanity and the natural world" all point towards a
philosophical dimension that forms part of their ecological
worldview.

An expanded circle of identity and care indicates who
these sustainability leaders are capable of identifying with

and caring about. Finding significant evidence of world-centric and planet-centric spans of care within this sample population of sustainability executives was a surprise. It was apparent during many of the interviews that the experiences of working on global poverty and ecological issues in many parts of the developing world had led to this expanded sense of identity.

Perhaps the most important of the five ways that sustainability leaders expressed ecocentric worldviews is enhanced systems consciousness. This is indicative of their capacity to see a wide range of interdependence not only with the Earth's ecosystems, but also internally within their organizations and externally across multiple cultures, ethnicities and countries.

Based on their roles in highly complex and large-scale organizations, these global sustainability executives appear to recognize the interconnectedness of social, economic, environmental and political forces influencing the sustainability initiatives they champion. In order to navigate the complex global challenges facing us, a strong and highly developed systems consciousness may be the most crucial capacity sustainability leaders of the future must possess. It is also indicative of what developmental psychologists categorize a post-conventional worldview, which is the subject of the next part of this book.

> In order to navigate the complex global challenges facing us, a strong and highly developed systems consciousness may be the most crucial capacity sustainability leaders of the future must possess

Part 3: How sustainability leaders think

Sustainability leaders are informed by an expanded view of how the complex universe operates. Being a sustainability leader means letting go of the ego-driven certainty of right answers and genuinely engaging with different points of view.

Mary Ferdig

8
Interior dimensions of leadership

Since World War II, researchers have been asking: "What makes great leadership?" "How do successful leaders become who they are?" and "What are the skills and capacities that make them successful?"

Until the middle of the 20th century, centralization of power and control were the primary themes in the leadership literature. As a result, many answers to these questions initially centred on the military, and many of the great leaders written about in the United States were famous generals such as Patton and Eisenhower. Based on principles of scientific management and the rational man, modern leadership theory began to emerge in the middle part of the 20th century.

Personality traits, intelligence and leadership styles became a primary focus for describing great leadership. Transactional and transformational theories of leadership that focused on the ability of leaders to motivate their workers also emerged in the 1950s and 1960s. Collectively, these theories became known as "great man" theories of leadership. Over the second half of the 20th century, leadership scholars have conducted more than a thousand studies in an attempt to determine the definitive styles, characteristics or personality traits of "great man" leaders.[65]

The major assumption underlying these "great man" theories is that leaders are born and not developed. Further, that these great leaders possess extraordinary capacities including charisma, intelligence and confidence.

Significant alternatives to the "great man" theories emerged in the late 20th century that focus more on self-awareness and broader purpose. Terms including "servant leadership", "primal leadership", "authentic leadership" and "enlightened leadership" began to appear.

Based on my experience in the corporate world and in higher education, the most important capacity for leaders to develop involves their psychological and emotional development. This includes a greater awareness of values, motivations, and deeper purpose. For this reason, I use *Servant Leadership*, *Primal* (emotionally intelligent) *Leadership*, and *The Fifth Discipline Fieldbook* as my primary course texts.

> The most important capacity for leaders to develop involves their psychological and emotional development

When my awareness of the ecological crisis deepened, I discovered how servant leadership and emotionally intelligent leadership lead to serving not only the people around us, but also the natural environment around us. I discovered new ways that these leadership philosophies can contribute to a psychology for sustainability leadership.

8.1 Servant leadership

By proposing that the real work of a leader is to ask themselves the question, "How can I best use myself to serve?", former AT&T executive Robert Greenleaf proposed a new type of leadership philosophy focused on service to others instead of ego and power. In doing so, he catalysed a family of more holistic leadership theory based on interior, psychological and spiritual dimensions.[66]

Greenleaf's book was a game-changer for me. When I first read it I had been leading companies for almost 20 years, or so I thought I had. Having been president of a pair of companies that were rolled up into a public company in the late 1990s, I experienced financial success. However, what was missing was a more expansive social and ecological

> Greenleaf's book was a game-changer for me

perspective. Possessing a highly competitive nature, I was the typical "Type A" executive. I defined success primarily on financial terms and drove those around me too hard. I experienced extreme amounts of stress and anxiety. During the evening and on weekends, I was unable to turn off and focus on being fully present with my wife and three young children. My approach to leadership was clearly not sustainable for the long term.

However, when I first read Greenleaf's words, I immediately understood that the real purpose of leadership, especially if it is to have any lasting value, is about service. I was able to rethink how I thought about the purpose of business. I began to experiment with ways I could empower those around me by seeking their ideas along the way from a broader perspective. I began to think more about the consequences of my actions and the role my company was playing in the world. I found that I no longer had to lead in a top-down hierarchical way, and that I could trust people to find new creative solutions that led to longer-lasting results.

> Servant leadership includes service to all species and the broader ecosystems within which business and our entire way of life exists

This experience ultimately led me to the realization that a philosophy of servant leadership can underlie a psychology for sustainability leadership. When seen through the lens of an ecological worldview, servant leadership expands to include service to all species and the broader ecosystems within which business and our entire way of life exists.

8.2 Emotional intelligence

The second major approach to leadership I use to
help students explore their deeper motivations for
sustainability leadership is rooted in research about the
emotional intelligence of leaders. In *Primal Leadership*,
psychologist Daniel Goleman and his colleagues Richard
Boyatzis and Annie McKee describe how great leaders
act foremost as emotional guides. Based on decades of
research from numerous sources, they highlight empathy
and self-awareness as vital capacities for transformational
leadership.[67]

A distinctive aspect of their research comes from the
field of neuroscience, where scientists are able to study
the limbic emotional centres of our brains. Neurosciences
researchers have been able to observe that when we work
together in groups, our emotions are part of what they call
open loop systems. This means that we effectively catch
feelings from each other. As we all know from our own
experiences in organizations, the emotions of leaders play
a disproportionately important role. Goleman and his
colleagues observe how the emotions of top leaders have
a sort of domino effect that ripples through a company's
emotional climate.

They conclude that in order to inspire deeper passion and
motivation over the long term, leaders need to create what
they call resonance in their organizations. They observe that
emotionally intelligent leaders have the unique capacity to
create a shared way of interpreting and making sense of any

given situation confronting their organizations, especially in times of chaos and crises.

As we continue to explore the interior dimensions of sustainability leadership, we can see that, in order to successfully lead organizations towards a deeper transformation to sustainability, a high level of emotional intelligence will be required. Goleman and his co-authors observe, "throughout history and in cultures everywhere, the leader in any human group has been the one to whom others

> Integrating emotional intelligence will have increasing relevance for sustainability leadership

look for assurance and clarity when facing uncertainty or threat."[68] Given our accelerating vulnerability to the effects of climate change and the inevitable emotional toll on people in organizations, integrating emotional intelligence will have increasing relevance for sustainability leadership. Later on, in Chapter 12, we will explore further the applicability of emotional intelligence to sustainability education.

8.3 Systems thinking

During leadership and strategy courses, I try various approaches to embed systems thinking in the minds of business students. My primary course text for teaching systems thinking is *The Fifth Discipline Fieldbook* by MIT professor Peter Senge and his colleagues at the Society for Organizational Learning. I consider the *Fieldbook* to be one of the best resources available to introduce business students to expanded ways of thinking.

Explaining that at its core, every organization is a function of how its members think and interact, the authors include various strategies to expand the capacity for systems thinking such as personal mastery, ladders of inference and mental models. All of these are designed to help people carefully examine and change the way they think.[69]

What I find noteworthy is that, by using permaculture to teach principles of natural capitalism, business students appear to grasp systems thinking in a deeper way. Then, as they complete personal sustainability projects based on alternative transportation, waste repurposing, renewable energy, green building, water stewardship and carbon reduction, students make further connections. During an 11-week term, a new understanding of systems thinking and interdependence between healthy ecosystems, human health, human rights and economics stabilizes in the minds of students.

> By using permaculture to teach principles of natural capitalism, business students appear to grasp systems thinking in a deeper way

8.4 Is sustainability driving human development?

In the last decade, much has been written about sustainability leadership from the perspective of innovation, strategy, marketing and management.

However, except for a small number of recent small-sample studies and theoretical articles, no cohesive theory of sustainability leadership has reached the corporate or academic mainstream literature. For instance, in the most recent edition of *Leadership: Theory and Practice* by Robert Northouse, one of the most widely used leadership texts in university courses, there is no specific mention of sustainability leadership among its chapters, much less any cohesive theory.[70]

> No cohesive theory of sustainability leadership has reached the corporate or academic mainstream literature

Based on my observations about systems thinking emerging in the minds of students while studying permaculture, a new pair of questions emerged as part of my research into worldviews. How does the practice of sustainability affect worldviews? And vice versa, how do worldviews affect sustainability leadership? In other words, what is the relationship between sustainability and the process of human development? A way to explore these questions presents itself when we consider that developmental psychology researchers have empirically demonstrated over several decades that enhanced systems thinking is a key indicator of advanced stages of psychological development.

9
Human development

Although there are early references to human development in ancient religion and philosophy, the modern concept of development of self through differentiated stages was pioneered by the Swiss child psychologist Jean Piaget in the middle of the 20th century.[71] In the 1960s, psychologists including Erik Erikson, Lawrence Kohlberg, Jane Loevinger and others expanded the study of childhood stage development to adults.[72]

In the early 1980s, Harvard psychologist Robert Kegan contributed the term "constructive-developmental" to the psychological literature. Since then, developmental researchers including Boston College professor emeritus William Torbert, Suzanne Cook-Greuter, philosopher Ken Wilber and a new generation of scholars from the worldwide integral community including Terri O'Fallon, Sean

> Significant new contributions have been made to our understanding of how adults construct knowledge about the world around them

Esbjörn-Hargens, Barrett C. Brown and many others have made significant new contributions to our understanding of how adults construct knowledge about the world around them through specific hierarchical stages of increasing complexity across their life-span.

Research psychologist Suzanne Cook-Greuter describes these stages as the recognizable stories human beings tell about who they are, what is important to them and where they are going. She explains that these stages are coherent systems of how people make meaning of their lives and that these systems evolve in complexity throughout one's life. She concludes that this is at the heart of what drives human beings.[73]

Collectively, developmental researchers have shown that there is a pattern to the stages of consciousness that adults can potentially traverse over the course of their lives. Research further shows that each stage can only be reached by journeying through an earlier one. Once a stage has been realized, it becomes a permanent part of how an individual interprets the world around them. Most importantly, as pertains to how sustainability is interpreted by diverse groups of people, each stage affects what an individual can be aware of, reflect on and act on.

> Most importantly, as pertains to how sustainability is interpreted by diverse groups of people, each stage affects what an individual can be aware of, reflect on and act on

As part of the theoretical framework for my research, I became interested in a broad stage of development known as "post-conventional" where the capacity to collaborate across boundaries and see greater interdependence of systems becomes a more consistent way of interpreting and acting in the world.[74]

9.1 Post-conventional worldviews

In his work on moral development, Lawrence Kohlberg first used the terms "conventional" and "post-conventional" to refer to two broad stages of moral development. Since then, developmental researchers have subdivided stages of development into three levels known as pre-conventional, conventional and post-conventional. Leadership consultant and human development researcher Barrett C. Brown explains that "Worldviews change over time, becoming more complex and encompassing ... After decades of research in the areas of cognition, morals, values, ego development, and other facets of human nature, it is clear that there are at least three general stages of development: pre-conventional, conventional, and post-conventional."[75]

Pre-conventional stages are associated with impulsive, opportunistic, and lower levels of psychological maturity. Conventional stages are characterized by conformance with social conventions, achieving expertise and efficiency, and short-term economic goals. Post-conventional stages are characterized by reframing problems with a deeper

> Post-conventional stages are characterized by reframing problems with a deeper understanding of context and interdependence of systems

Conventional worldviews	Post-conventional worldviews
Short-term economic goals and maintaining power are highest priorities	More aware of social and environmental implications over longer time-frames
Primarily focused on achieving efficiency when decision-making and interacting	More collaborative, relational and process-oriented when decision-making and interacting
Sees problems and challenges primarily in black-and-white terms	More aware of complexity and interdependence of systems
Maintains loyalty to group; success is defined by organization	More aware of diverse contexts, cultures, worldviews and multiple consequences
Heavily concerned with conformance to group norms and social conventions	More willing to challenge group norms; seeks to transform systems and organizations
Approach to problem-solving primarily using data, logic and scientific rationale	More aware of emotions, unintended consequences; uses intuition
Accepts feedback only from objective sources and superiors	Consistently inquiring and open to feedback from diverse sources

TABLE 9.1   Description of conventional vs. post-conventional worldviews
Note: Table adapted from Torbert, 2004; Kegan, 1994; Boiral et al., 2009; Brown, 2012; Esbjörn-Hargens and Zimmerman, 2009.

understanding of context, interdependence of systems, and a greater awareness of environmental and social implications over longer time-frames.

Table 9.1 provides a brief description of several of the most important distinctions between conventional and post-conventional worldviews.

9.2 Action logics

In the domain of corporate leadership, William Torbert and his colleagues at Action Inquiry Associates, including Elaine Herdman-Barker, Hillary Bradbury Huang, Dana

Carman and Nancy Wallis, have conducted extensive field research designed to measure scientifically the stages of development in the context of organizational leadership.[76] As part of their multi-decade research exploring stages of development and their implications for leadership, they have methodically created and refined descriptive labels for each stage, called "action logics". The action logic descriptions capture key characteristics of each stage that become the frame through which people translate their thoughts, feelings and perceptions into actions. Highlighting the relevance for sustainability leadership, David Rook and William Torbert explain that, "What differentiates leaders is not so much their philosophy of leadership, their personality, or their style of management. Rather, it's the internal action logic— how they interpret their surroundings and react when their power or safety is challenged."[77]

In addition to the term "action logics", developmental researchers have periodically used the terms "centre of gravity", "cognitive map", "mind-set", "mental pattern", "meaning-making system", "order of consciousness" and, most broadly, "worldview" to define stages of development. For the purposes of this book, I have chosen to use the term "worldview" for the two broad stages of development described above. I have done this in order to align with my use of the term "ecological worldview" used throughout the book.

Post-conventional worldviews therefore refer to the more advanced stages of adult development. Based on decades of research in the corporate domain, Torbert, Kegan, Cook-Greuter and numerous other developmental researchers have

shown how leaders with post-conventional worldviews have
enhanced capacities to transform organizations.[78]

9.3 Ecological crisis and longer life-spans driving human development

Based on many years of research in counselling, education
and management, Harvard psychologist Robert Kegan and
his colleagues created a stage model of human development
they characterize as "orders of consciousness". Closely
related to Torbert's model of action logics, Kegan's model
also offers specific descriptions of the capacities involved in
each of his five orders of consciousness. Post-conventional
worldviews emerge approximately at the level of fourth-
order consciousness. In his widely acclaimed 1994 book
In Over Our Heads, Kegan explains that this involves
developing a deeper, internal set of convictions that can
serve as a set of larger visions or values. This larger set of
visions or values then allows one to take a broader, longer-
term perspective and have the capacity to reframe a given
situation, especially when under stress.[79]

Referring to the mental demands placed on us by the
complexity of the modern world, Kegan described the
human predicament as "over our heads in a sea of icebergs".
Considering how much less information about the ecological
crisis was available when he wrote this in 1994, I surmise
that he was not using the metaphor with climate change in
mind. Given the ways that climate change and its associated

environmental problems have dramatically accelerated this complexity over the 20 years since he wrote those words, ironically his metaphor could not be more prescient.

With his shivering image of man and nature, Kegan unintentionally captured then what we now know today. Global sustainability leaders are navigating quite literally through "a sea of icebergs". To support them, we need to develop a new way of thinking that is up to that task. The work of Kegan and his colleagues continues to shed light on this new way of thinking.

> Global sustainability leaders are navigating quite literally through "a sea of icebergs"

Recently Kegan invited us to consider that, as a species, we are collectively trying to figure out the answer to the problem that we have devised the means to our own demise. His big idea is that, based on his research that people generally reach post-conventional fourth-order consciousness towards the later part of their lives, and given the dramatic increase in our life-spans today, there are now many millions more of us that are going to reach fourth- and fifth-order consciousness in the coming decades. He describes this as an evolutionary process. By doing so, he offers a new intriguing link between the relevance of post-conventional worldviews and sustainability leadership development in the coming decades.[80]

> A new intriguing link between the relevance of post-conventional worldviews and sustainability leadership development in the coming decades

9.4 Emerging research into the psychology of sustainability leadership

Responding to the knowledge gap that currently exists about the psychology of sustainability leadership in the corporate world, there have been several recent theoretical and small-scale empirical studies thus far that use developmental psychology to explore interior dimensions of sustainability leadership.

A robust theoretical study found in the sustainability leadership literature is by management scholar Oliver Boiral and his colleagues at the University of Montreal. These scholars observe that the academic literature often incorrectly assumes that sustainability leaders share the same values and ways of approaching environmental challenges. With a focus on sustainability leadership behaviour, the authors use the action logic descriptions to hypothesize specific potential behaviours of leaders at each stage of development.

Boiral and his colleagues conclude that the characteristics associated with post-conventional action logics or worldviews are most closely aligned with the demands of sustainability leadership. They observe further that post-conventional consciousness equips individuals to navigate the complexity of sustainability issues and mobilize individuals for environmental causes within organizations. However, they acknowledge that the absence of ecological worldviews in the developmental research is due in part to the psychometric instruments being developed in an era when ecological issues were less salient.[81]

In terms of empirical research, an exciting and generative series of empirical studies have been conducted in recent years. These include Wayne Visser and Andrew Crane's study of sustainability change agent archetypes,[82] Abigail Lynam's study of sustainability education graduate students and faculty,[83] and Simon Divecha and Barrett C. Brown's study of action logics and integral sustainability.[84] Most closely pertaining to my research, there are three recent studies that include a more explicit focus on ecological worldviews in the context of corporate leadership.

In the study I describe earlier in Chapter 6, sustainability researcher Katrina Rogers explored the connection between individual worldviews of executives in a European multinational and their ability to confront global environmental challenges. Using the ecological self as a theoretical lens, she found that executives were able to identify specific moments that led to a different way of thinking about the environment. While certain executives characterized these changes as epiphanies, others described a more gradual evolutionary shift. All of the executives reflected on these shifts as being a permanent change in the way they conducted their professional lives.

Exploring the connection between ecological self and worldviews, Rogers reported that executives that experienced these behavioural changes demonstrated worldviews that appeared to be more advanced in several specific ways. She found that these executives demonstrated a more highly developed sense of complexity, systems thinking and interdependence. She speculated that further use of the ecological self, and development of a new instrument, could lead to new insights and a deeper understanding about how leaders develop advanced capacities to confront the global environmental challenges.[85]

In a second recent exploratory study described in her book *Big Bang Being*, psychologist Isabel Rimanoczy interviewed 16 corporate leaders that were actively involved with sustainability initiatives in order to learn what inspired their focus on sustainability. She found that the majority of her research participants developed what she characterizes as a social sensitivity and a sense of personal mission.[86]

In an interesting connection with the experiences that shape ecological worldviews I report in Chapter 4, she recounts how the sustainability leaders she interviewed often reflected on specific experiences at specific moments in their lives. These moments occurred in their early childhoods through their upbringing and role models or by a type of awakening or meaningful encounter as adults. Among them is the epiphany Ray Anderson describes as a "spear through the chest" when he realized that the economy is embedded in nature.[87]

> These moments occurred in their early childhoods through their upbringing and role models or by a type of awakening or meaningful encounter as adults

Rimanoczy reports that in many of the interviews she conducted, participants described that an important element of their mind-set was a deep closeness to nature. These sustainability leaders spoke of the importance of paying close attention and observing the wisdom of nature. Further, the sustainability leaders reported having transformational experiences connected to encounters with nature, which became a key shaper of their mind-set.

Rimanoczy also explores resistance to sustainability. She identifies numerous core values of Western culture that she attributes to this resistance. Among these are economic growth, independence and control. In her discussion about control, Rimanoczy describes the impact of anthropocentric worldviews and how they lead to a wide range of activities that cause environmental degradation, including mining, industrial fishing, oil exploitation, industrial farming, genetically modified crops, rainforest clearing, and the dumping of chemical waste into sources of fresh water.

Observing that our belief in the superiority of humans over nature is evident in our widespread destruction of ecological systems, Rimanoczy concludes that our anthropocentric worldview is leading to what many are calling the sixth mass extinction and that the alarm bell is ringing loudly.

In a third recent study into how sustainability leaders with later stage action logics design sustainability initiatives, leadership consultant and human development researcher Barrett C. Brown illuminates ways that leaders think and behave with respect to sustainability-oriented change initiatives. Based on his interviews with a small group of sustainability leaders across various sectors, he reports that they design from a deep inner foundation, use systems thinking, access non-rational ways of knowing and adaptively manage through dialogue with the system.

Brown makes the observation that the discipline of sustainability leadership is still in the very early stages and the lack of large-scale, empirical research currently makes it difficult to draw specific conclusions about the nature of sustainability leadership. He points out that our understanding of the psychology of sustainability leadership is therefore extremely limited and concludes that the more

> The more we understand what drives the behaviours required to lead sustainability initiatives, the easier it will be to cultivate them

we understand what drives the behaviours required to lead sustainability initiatives, the easier it will be to cultivate them.[88]

Based on the potential for my interviews with sustainability executives to provide additional insights through the lens of developmental psychology, a new research question informed the next stage of analysis:

Do the descriptions by sustainability leaders of their ecological worldviews and their motivations for their work in sustainability suggest a more advanced stage of psychological development?

Guided by this question, I went back through my interview transcripts looking for evidence of post-conventional worldviews. In the next chapter, I describe what I found.

10
Expressions of post-conventional worldviews

By using the descriptions of post-conventional worldviews as a deductive lens for further analysis of the transcripts, I uncovered a surprising amount of evidence suggestive of post-conventional worldviews. Overall, I was able to identify five distinct ways sustainability executives expressed post-conventional worldviews during our interviews. A complete description of my thematic analysis methodology is provided in Appendix B.

10.1 A greater awareness of context and diversity of worldviews

This first set of quotations suggest that global sustainability leaders appear to have developed an enhanced capacity to see a growing range of worldviews and contexts as part of their global sustainability practice. This capacity, and the capacity to be aware of one's own worldview and context, is one of the most crucial differences between conventional and post-conventional worldviews. A range of social, environmental, economic, cultural and political worldviews were described in various ways by executives during their interviews.

Their capacity to recognize and appreciate various worldviews and contexts appeared to be further magnified by the global scope of their multinational corporations. There were many instances where it appeared that the inherently complex and interdisciplinary nature of their work in global sustainability was forcing them to develop this capacity. The diversity of worldviews and contexts exists both internally within their organizations and externally with a wide range of global stakeholders.

For example, this executive, the chief sustainability officer at a global consumer products company, described his work as follows:

> In my role as CSO (chief sustainability officer) everybody has an opinion because the idea of sustainability is so personal. When I come into a meeting with groups of our employees I don't know what their individual priorities are. One person cares about homeless people, another about animal rights, and other about climate change, and some of them don't care at all. In my job, everything is framed about caring. So everyone cares about different things.

A second chief sustainability officer, this one at a global travel services firm, described how she thinks about the diversity of worldviews in the following manner:

> When you think about sustainability you have to think about both ends of the spectrum. It's about bringing everyone to the centre from both sides, from the non-believer to the eco-enthusiast. To me it's equally important that the eco-purist understands that there are dollars involved. Both viewpoints have to be brought in. When I sat down with folks from my finance department and they saw how smart managing energy consumption was they began to understand. Changing behaviours just for the sake of the environment does not resonate with everyone.

A third senior executive talked about how his awareness of diverse worldviews affected his approach to communication:

> Part of the reason that I don't stand on a soapbox is to allow a constant drip of improvements that starts to take root, one by one. I don't ask employees how they regard *our* efforts. I want my conversations with my team to be about *them*, and *their* needs … It's my experience that I regard these issues different than most people, I don't want to environmentalize as a religion. I'm motivated on a very practical level. There's a risk if I were to be seen as too extreme in my views. It's kind of a circuitous pattern.

A strong awareness of cultural context was apparent in many of the interviews. Considering that so many of the multinational sustainability executives interviewed have lived and worked abroad for many years, this makes sense. For example, the following four chief sustainability officers referred to their experiences in India, Sweden, Holland, China and Germany, and described how this has affected their thinking. This sustainability executive reflected on a recent work trip to India and how it gave her a new perspective:

> One thing that was very illuminating for me was my trip to India. There you have a lot of different environmental issues. Although they have the equivalent of a Clean Air Act and Clean Water Act, I don't sense it is very well enforced. So although they do have those policies, just the sheer magnitude of people and processes and outsourcing of our manufacturing and everything else in India, you really see all of that boiling there. It was very illuminating because before that I thought that things could be so much better, but not really acknowledging how far we've come. And then you go to India and it makes you realize these are real serious issues. The air quality alone is a major issue in terms of particulate matter and the amount of biomass that's being burned. You blow your nose at the end of the day and it's just black. So I think this was a good reality check for me.

A second chief sustainability officer at a major global consumer products company, who had worked for many years in Sweden, reflected on how this may have prepared him for his work in sustainability as follows:

> I was based in Sweden, which has a more collaborative culture. I think this prepared me for my work and sustainability. To get the sustainability agenda to work internally you have to have strong influencing skills. When I think about sustainability leadership in contrast to the 20th century "great man" style of leadership, you're more the collaborator-in-chief. You're not command-and-control, you're not omnipotent. I knew immediately that I did not know everything there was to know about sustainability. My background was in marketing. I know that I will never be an expert. It enforces a new level of humility because I know deep down that I need to collaborate with others in a more flexible way. I'm very grateful that I went into sustainability and feel very lucky because it is helping me develop some right-brain skills that I did not know had gone dormant. Practising sustainability has allowed me to collaborate at a higher level.

A third executive shared her perspective about the Dutch culture and relationship to technology:

> I lived in Holland for two years. Culturally the Dutch are so interesting. In regards to sustainability, I would describe them as being very techno-centric. They have this deep understanding

about why they're so vulnerable to climate change because of the fact that so much of the land is below sea level already. They have this intricate system of managing their agricultural land and the North Sea using these dyke and levee systems. They know that they would just be inundated creating a huge loss of agricultural land. However, through their technology they believe that they will overcome the environmental factor.

Lastly, demonstrating an awareness of cultural context, this senior sustainability executive shared how his time in China and Germany affected his worldview and understanding of global environmental geopolitics.

I've led environmental research labs in China and Germany. Running those groups for around seven years gave me a great perspective on environmental science, technology and culture. It sort of drove a stake in the ground for me. The Chinese know that they are running out of concrete, copper, steel and minerals. The best place to get them is in Africa. Unfortunately, they will do business with corrupt governments. They will go to the Congo no problem. For the Olympics they ran out of concrete.

These last four executives shared general observations about the importance of an awareness of worldviews and contexts in general, especially for global companies.

What I do think is the case is that if you are a top-tier person moving towards the C-suite you need to understand the broader environmental and social context to be successful.

I've had an evolving view. Twenty years ago I was involved with environmental activities. Working in management you're really working on a very different side of the conversation. I realized that there are all these different perspectives and challenges involved with addressing the environment. And there needs to be a middle ground.

If you think about the influence of technology, we all need to be aware of how what we are doing will affect the broader society. It is now an expectation that we do work that benefits developing countries and that we can offer more solutions around the world.

We're finding that more and more engineers are in tune with societal issues on a global level. For instance, we have many new healthcare and energy products that are designed for rural markets in India and China. It's really important that engineers have a cultural sensitivity to the emerging markets. We need enlightened engineers instead of those with a Western mind-set.

This last participant, a sustainability executive in the supply chain group of one of the world's largest computer companies, states that:

I've found that there are two types of people. People that are very interested and excited about sustainability and people who see it as just one more mandate from corporate that distracts them from doing their primary work, which is meeting their financial goals.

10.2 Holding longer historical and future time-horizons

The second set of interview quotations indicate a capacity to consistently think in longer time-horizons, another strong indicator of post-conventional worldviews. Many of the executives appear to have developed the mental pattern of putting sustainability in a long-term historical and future context, even while operating under the persistent short-term pressure of public companies.

In particular, developmental researchers have found that leaders with post-conventional worldviews think in decades, or in terms of generations or their own lifetimes. This can mean consistently thinking about consequences and systems in terms of multiple decades into the past and future instead of weeks, months or years.

Many of their reflections during the interviews showed evidence of this capacity. It was apparent that many of the participants were aware of the historical context within which they are working as well as the reality that it will take decades to accomplish their goals. In particular, the time-period of the early 1970s, when the modern environmental movement was born, forward to the middle of the 21st century was often referred to during my interviews. Many of the sustainability leaders interviewed possess direct and intimate knowledge of important events that took place in both the public and private sectors over the last 30 years and take these into consideration when thinking about initiatives related to sustainability.

By emphasizing their sense of history and long-term view of the future, the following quotations serve to illustrate this aspect of a post-conventional worldview.

> We can see how it will play out over the next ten years. I can see that if we do this right, we can really make change happen. It's so much about common sense. Why do we throw all this stuff away and create all this waste? Ten years from now there will not be green companies. Every company will have a set of best practices around sustainability. I won't need to show you the difference between green companies and other companies. The first step was getting these protocols in place. There was a tremendous amount of research that went into this. This will be the standard in ten years.

> I think people need to understand how everything fits in and relates to the future. What you're doing today affects what is going to happen ten years from now. What has changed is that now we ask ourselves: What do we need to do by 2050 as relates to climate change and carbon parts per million? What is the right thing to do? How should we be gearing the business to make that happen? How should we engage in legislation?

> What is the science behind this? What is going to
> be possible? Our plan has become how we can
> have all of our products put together in a way that
> allows us to have flexibility taking into account the
> large changes that will take place in the coming
> decades.

> In the post-industrial world we will see the result
> of our work in sustainability.

> All of us need to be doing things that are very long
> term and very visionary.

> What we've found in regards to our sustainability
> initiatives is that people go through phases of
> understanding over a long period of years. We
> continue to try to classify where we are on the
> journey.

During the interviews, climate change and the role of big
business were frequently put into a longer-term historical
and future context by numerous participants. Many of the
research participants represented their companies in a variety
of public–private–NGO coalitions related to climate change
and sustainability. Given their ecological worldviews, many
of these individuals spoke in very personal terms about the
issue of climate change and frequently put their thoughts
into historical context.

For example, this participant, who I quoted earlier as an
example of enhanced systems consciousness, said:

> My personal view is that we've got to find a way
> to move from the goal of just understanding the

natural environment to the realization that we ourselves are causing the environment to change drastically around us for the first time in the history of man. I think that changes the game. I think that I struggle with this personally. I don't believe that it is possible at this point to mitigate the damage we have started. Now we need to mitigate how bad it's going to be how we're going to help people deal with these environmental realities in the future.

An additional way that several executives used to talk about sustainability in a longer-term historical and future context was through the role of technology. The following three quotations illustrate this idea:

It has developed and blossomed over the last ten years. Certainly when you look at environmental and social challenges that we face, we went from a risk to an opportunity shift. People's approach in solving problems shifted to redesigning instead of putting band-aids on things. Instead of putting all our energy into cleaning up our environmental mistakes, let's redesign our manufacturing processes instead to prevent it from happening in the first place.

Engineers are going to have to be retrained so that they understand their true impacts on materials use, stewardship, etc. It's a huge problem. It's a bigger challenge than putting a man on the moon. They have ridden that horse as far as the horse can go. The horse can take us to Cape Canaveral.

> We need a different vehicle to get us to the moon. We need to make a quantum leap here. We need a new, larger, framework that is bigger than an incremental optimization. You do need the leader to stand up and do what Kennedy did. A lot of people don't know that Kennedy originally wanted to go to Mars. Getting to the moon was really something. We need a similar approach.
>
> Sustainability is all about connecting lots of people and information and making that information available in real time. We're working to create solutions across the supply chain and around the world to produce information in real time. As an enabler of sustainability this is still in its infancy.

This executive was one of the few in my study that was able to express the ideas of a steady-state economy that form the bedrock of a new ecological paradigm. His comments also demonstrate this capacity to think in a longer time-horizon:

> The longer-term issue that I see in terms of leadership is that eventually we have to deal with consumption with a big C, not just Cradle to Cradle thinking, but we have to transition to business models that do not depend on growth, that in fact thrive on the basis of reducing consumption. Right now the business world is getting on board with being more efficient because they know we are running out of natural resources or they are afraid of climate change or political

instability, they are on board with this first phase,
but not the second phase.

Finally, by citing the Chinese philosopher Lao Tsu and
Leonardo da Vinci during their interviews, these two
executives went back centuries. Reflecting on the sheer
diversity of issues and events, this participant, the CSO
from a large electronics company, quoted an ancient Chinese
proverb, "We live in interesting times".

When speaking about a recent engineering change related
to sustainable design, this participant recalled a trip to
Europe and the connection he made with his work around
sustainability and the famous Renaissance inventor:

> What I was delighted by was walking around
> Rome, Italy and coming across da Vinci's
> drawings of little machines that he had designed.
> There were some similarities to the concepts we
> had used. I realized that da Vinci had a sense of
> not sending waste out into the physical world.

10.3 The self in the system

As a consequence of being able to see a greater diversity of
contexts and a capacity to think in longer time-horizons,
an awareness of the broader system emerges. The next
set of quotations demonstrates an enhanced systems
consciousness, which is a third crucial characteristic of
post-conventional worldviews. A highly developed systems

consciousness is also a core component of the ecocentric worldviews we saw in Chapter 7.

Based on their roles in highly complex and large-scale organizations, these global sustainability executives appear to recognize the interconnectedness of the social, economic, environmental and political forces that influence the sustainability initiatives they are responsible for championing. This enables them to see a wider range of interdependence both internally within their organizations and externally across multiple stakeholders, countries, cultures and ecosystems.

They become less quick to assert their own worldviews while maintaining an enhanced understanding that they cannot control all the variables within the system.[89] They also begin to see themselves as part of systems within systems. Developmental researchers have called this the capacity to see "the self in the system".[90]

For example, this senior sustainability executive at a major apparel and footwear manufacturer offered this reflection that demonstrates her systems consciousness. You'll note that I included the first part of this quotation in Chapter 7 as it demonstrates an ecocentric worldview as well. Here we begin to see the commonalities and relationships between ecocentric and post-conventional worldviews:

> I think probably that where I come from in terms
> of my ecological worldview is systems thinking
> and the interconnection of so much of what we
> do and our impact on the environment. I've spent
> a lot of time over the years around sustainability

and been exposed to a lot of what's going on
in the world. I have a science background. I did
my masters in psychology but I started out with
biological sciences. My interest in the science
side of this that just really been just help shape
that. I've always understood at the fundamental
basis that the economy and society are within
the context of the environment. So we really
can't do anything without paying attention to the
ecological context within which we live. That's
probably the worldview piece.

Here are three more quotations from senior sustainability
executives I included in Chapter 7 that demonstrate both
ecocentric and a post-conventional worldview through an
enhanced systems perspective:

The next circle out there is the whole planet …
Quite often it breaks down to understanding
yourself and your dependence on nature. There's
an interrelationship obviously. It means taking
yourself and your team out into the world and
becoming aware of how you are impacting the
bigger ecosystems and making linkages. You could
scale that down or up however you want to. But
it's basically how the social and economic systems
of the human community are in relationship with
ecosystems. You need to have an understanding
of all that kind of stuff in order to be skilfully
engaged.

When you step into a role like this what you
think will inspire you changes. For example, I

never thought I would be so excited about trash. However, I realized that I was getting excited about systems thinking. In order to be a real change agent you have to understand the whole system. One day I put on my gloves and went through the trash in one of our buildings. When I thought about waste diversion, I began seeing the entire global waste system.

First, that the more you work on sustainability you realize it is not just connected to other issues, but the same as other issues, like: ethics, religion, business, family, education, health, poverty, respect, government.

This sustainability executive and major multinational food company, who I also quoted in Chapter 7, described how she and her team see their company as part of the global food system. She described to me how they're using their company as a platform to influence other multinational companies towards more sustainable agricultural practices. This represents yet another expression of an ecocentric and a post-conventional worldview.

We have a series of position papers that we worked on about a broad range of topics ... From climate disruption to local food systems and the value in terms of big world/small planet point of view ... the finished papers will be external. They are very good papers for conversations. We are using the papers to influence other multinational companies who may not understand what we are

> doing … It's very important to us for our peer
> companies to get what we are doing. They don't
> understand what we are about. So the goal is to
> give them transparency about what our beliefs
> are and what our ambitions are. The papers are
> country specific. The French are very progressive.
> No genetically modified food. In the next several
> years we can be a catalyst for safe food. If we
> capitalize on our ability to tell our stories and to
> really engage our consumers, that is where our
> strength comes.

This executive, the head of environmental strategy at a
global technology company, became very animated when
talking about the long-term potential of scaling up the
solutions developed within his organization to address wide
variety of global challenges.

> I would say that what makes this job most
> compelling for me is that, despite all these really
> cool things that we are doing, we have not even
> scratched the surface in terms of what type
> of unique contributions we can make systems
> wide … galvanizing people and shifting from a
> "profit margin and cogs" type of company to
> where we are helping to address societal issues at
> a scale that we can't fathom and unlocking our
> potential.

This participant, the CEO of a major multinational
company, also reflected on scale in the following manner.

> I think that we understand now that scale is a tool
> and you can use scale to do good things or use
> scale to do bad things. We try to use scale to do
> good things.

This final quotation also serves to illustrate the systems
thinking aspect of a post-conventional worldview:

> It comes from learning where we had a role in
> impacting the environment, not just the planet
> itself but also the people working for the company.
> Finding an appropriate balance between all three
> areas: economic, social and environmental. I
> carry the bulk of the environmental focus but the
> foundation is social justice. I struggle with the
> balance and how to involve everyone. I carry it as
> a personal mission.

10.4 A widening circle of identity and care

The capacity to identify with a widening circle of human
communities, the biosphere and all species is another
important characteristic of a post-conventional worldview.
This is also referred to as "span of care" by developmental
researchers. This is significant for a psychology of
sustainability leadership because expanding who and what
we *identify* with expands who and what we *care* about
and eventually *act* on. This is where we find leaders at the
intersection of ecocentric worldviews and post-conventional
worldviews, which poses interesting questions for future

research. For this reason, I repeat several of the quotations from Chapter 7 here.

Developmental researchers have found that the worldviews of leaders with post-conventional outlooks can become wider until they embrace everyone on the planet as well as the biosphere itself, as philosopher Ken Wilber has noted.[91] Finding ample evidence of world-centric and planet-centric spans of care within this sample population of sustainability executives was a surprise. It was apparent during many of the interviews that the experiences of working on global poverty and ecological issues in many parts of the developing world had led to this expanded sense of identity and care.

When reflecting on their work, there were numerous instances where their words indicated a strong felt sense of the entire global community. Once again, given the global reach of their companies, this comes as no surprise. However, I was struck by how often a global perspective was mentioned during my interviews.

For instance, when reflecting on the issue of climate change, these participants highlighted a missing perspective from the political climate change debate:

> Of course from a global perspective climate change is an enormous issue that we should be addressing. But I think one thing that is a little bit absent from these conversations is the outsourcing of our industrial processes to these other countries and our being ignorant of the effect of this. Also the reality that there is so much of just basic environmental protection that is not happening in developing countries. I think

that's under-reported. Most people know there's
pollution in China and in India, but we haven't
included that within our global environmental
goals as well as I think it could be.

This executive spoke of climate change being an issue of
equity for people in underdeveloped countries throughout
the world:

Ultimately climate is an issue of ethics and equity,
and solving it seems like an obligation to our kids
but also to poorer people around the world.

This next participant, a long-time senior sustainability
executive at a major apparel manufacturer, reflected that our
language and culture are still embedded in our patriarchal
society. By then making the observation that sustainability
could benefit from the language and culture of a matriarchal
perspective, she demonstrates how an expanding circle of
care can come out of the feminine:

I think we are honing our approach. It's an ever-
widening circle of learning. The work we're doing
on diversity and culture, recognizing what are
the patterns and the artefacts in the culture, I
continue to find it so helpful ... How do we lead
going forward? We need to move towards a more
matriarchal society from the dominant patriarchal
societies. We continue to need to see that
sustainability is embedded in the patriarchal and
explore how we can move into a more feminine
nontraditional approach.

This executive was able to articulate a specific point in time where his circle of identity expanded from himself as an organizational leader of other human beings to also serving ecosystems.

> It was there that I realized I made a shift from being primarily interested in my own experience of being a leader and interpreter to actually understanding ecosystems better in order to be of service and in some way conserve or restore ecosystems.

This last quotation from a sustainability executive from the coffee industry demonstrates his circle of identity as a global citizen. In it he reflects his compassion for people living in poverty in Nicaragua and how this has led to the evolution of his company's corporate responsibility initiatives:

> For many years I've been involved with fair trade and corporate social initiatives in Central America. Traditionally, we work with the co-operatives and the larger growers on pricing and to help improve their operations and the labour conditions for their workers. About five years ago as part of a trip to Guatemala I sat down with a small group of people in one of the villages to get a sense of what things really looked like from their perspective ... We were after some new perspectives so one of the things we did is separate husband and wife in taking the survey. The questions were very broad from how much

land people farmed, to how it was being planted,
to the details of their personal financial issues
including exactly where their income was coming
from and what the seasonal issues were for their
families.

10.5 A consistent capacity for inquiry

This final set quotations suggests that many of the
sustainability executives I interviewed are consistently able to
maintain an open mind and be in a place of inquiry, which is
yet another important characteristic of a post-conventional
worldview. This enables them to remain open and learn from
their environment. It allows them to listen more closely and
ask more generative questions, especially while under stress.

For example, this long-time senior sustainability executive
reflected:

> Earlier in my career, I had one sense of what
> environmentalism was and what approaches could
> be taken. In some ways, I was idealistic because
> I thought that if environmental education was
> framed in such a way, then anyone can learn
> and everything was possible ... I was framing
> everything as being beneficial and explaining
> why you want them to go in that direction. But
> then you ultimately don't have control over why
> or what everyone does. That said, you also don't
> really understand fully why people have done what
> they have done and how they may be constrained

> in some ways ... So lately over the last six or seven
> years I've come to understand why certain things
> are on the trajectory they are. It kind of just places
> the idealism inside the realism.

Reflecting generally on his frustration with the tendency of other employees in his company and in the general public to reduce things to more simple binaries, this senior sustainability executive put it:

> It's not as black and white as people see it ... It's
> kind of a theatre of the absurd and somewhere
> in-between ... it forces me to maintain a sort of
> confident humility.

Along the same lines but even more succinctly, this senior sustainability executive put it:

> I'm in a constant phase of rethinking.

Or this executive who spoke to me about how to communicate the essence of sustainability and wondered out loud:

> How do we put it to the heartbeat of the
> company?

This long-time sustainability executive, whom I had met at numerous conferences, shared this observation during our interview about needing to reframe sustainability and focus more on failures instead of successes:

> We as a business community are getting better at
> being less bad ... not yet being good. The pace

of change is nowhere where it needs to be. If we look at where we thought we would be now five years ago the progress has been very slow. I learn more talking to my colleagues about where they are failing. The most valuable lessons come from when we stumble. When I travel to conferences like CERES, BSR, etc., everybody wants to tell success stories. However, then we walk away in denial that although there are successes, we need much more honest and open talk about where we are failing.

The following quotation is from a senior consultant who facilitates a support network for chief sustainability officers. He spoke about the singular capacity for chief sustainability officers to *listen* on behalf of their organizations.

They all have unique positions within their company. Most of them are sort of the consciousness of their company. Someone has to be the listening post for what's next. They're sort of the Indian scouts for the company.

This chief sustainability officer at a global technology company reflected on how his background as a chemist allowed him to work on concepts he can't see:

For me I feel sustainability is too important and when you get down to it that's why I'm doing it. Being a chemist at a company, I know I don't have all the answers ... Chemists are really good at solving problems using the scientific method.

> We're good a working on concepts that we can't
> see.

This participant, the CEO of an iconic global consumer products company, described how sustainability has forced him to learn how get more out of his comfort zone.

> We started to do more ethnographies and talking
> to different kinds of people. It's really more about
> *their* world and *their* planet. You have to get out
> of your comfort zone and talk to people that you
> normally don't talk to.

This last quotation refers to what a vice president of sustainability at a global food company experienced during a retreat with the deep ecologist Joanna Macy. It shows that his vulnerability in the context of the ecological crisis is related to his capacity to ask deeper questions:

> I did a dense and deep work with Joanna
> Macy. Most of my peers don't want to be that
> vulnerable. The work I did with the sustainability
> consortium was great but I think it could've gotten
> deeper.

In Part 2 of this book we saw how an awareness of ecological embeddedness and planetary ecosystems, the belief in the intrinsic value of nature, enhanced systems thinking, and an expanded circle of identity are indicative of ecocentric worldviews and ecological self.

Here in Part 3, we have just seen how a greater awareness of context and diversity, thinking and longer time-horizons, seeing the self in the system, a widening circle identity, and a capacity for inquiry are indicative of post-conventional worldviews.

In the final part of the book, we see how together these types of worldview drive new collaborative approaches to leadership and explore specific ways they can be used to cultivate a new psychology for sustainability.

Part 4: The future of sustainability leadership

We all have the extraordinary coded within us, waiting to be released ... The ecological crisis is doing what no other crisis in history has ever done—challenging us to a realization of a new humanity.

Jean Houston

11
The collaborator-in-chief (with an ecological worldview)

In *The Necessary Revolution*, Peter Senge and his co-authors observe that:

> a profound shift has been unfolding for years, as countless collaborative networks have been established around a number of issues pertaining to sustainability ... Building the capacity to collaborate is hard work and demands the best of people, particularly when it involves people from different organizations with different goals and with little history of working together—maybe even with histories of distrust and antagonism.[92]

This new level of collaboration across boundaries is something Willamette University professor Elliot Maltz and I wrote about in the *Journal of Corporate Citizenship* in our 2013 article entitled "Cultivating shared value initiatives".[93] However, at the time I had yet to use developmental psychology and ecological worldviews as lenses to explore how sustainability leaders *think* about nature, leadership and collaboration.

In response to questions about the challenge of implementing large-scale sustainability initiatives and transforming the culture of their global companies, numerous executives I interviewed shared how their long-time work in sustainability was changing how they *think* about leadership. The majority of the executives I interviewed had at least a decade of sustainability leadership experience in the corporate world, with many of them having worked in the sustainability field for up to 20 years. Given what we now know about their deeper motivations, and considering their influential roles and the global scale of their companies, we can begin to perceive the hidden power of ecological and post-conventional worldviews.

Through further rounds of inductive analysis of the interview transcripts, I identified numerous expressions of highly collaborative approaches to leadership. One of the most prominent was the way they now perceived leadership as being a much more collective process. Key words and phrases such as "leading from the middle", "influencing without control", "collective wisdom", "from fear to trust", "both ends of the spectrum" and "away from the typical hierarchical approach" supported this finding.

For example, this participant, the chief sustainability officer at one of the largest computer companies in the world, shared her observation:

> I think sustainability has caused leadership to
> evolve within our company. When I joined the
> company the culture was really about seniority.
> Now it is more about leading from the middle.
> With 300,000 people there really is no other way.
> It would take a year to get the word out on our
> initiatives. You would be a dinosaur. With my old
> jobs around change management and diversity
> it makes it really difficult. What we do in the
> corporate function is very much trying to influence
> without control.

Along the same lines, this chief sustainability officer at a
global travel services firm, quoted previously in relation to
her awareness of diverse worldviews, offered this observation
about leadership:

> When you think about sustainability you have
> to think about *both ends of the spectrum*. It's
> about bringing everyone to the centre from both
> sides, from the non-believer to the eco-enthusiast.
> To me it's equally important that the eco-purist
> understands that there are dollars involved. Both
> viewpoints have to be brought in.

This chief sustainability officer at a major automobile
manufacturer offered a similar reflection about leadership:

> I think what comes up for me is that there has
> been a change in leadership over the years to use
> more of everyone's collective wisdom. When I
> came in there was a group doing sustainability
> but it was more on the side. There was a lot of

tension inside the company because of what they were doing. Most people felt that it was a side thing. When that CEO asked me to take over the sustainability function, he asked me to integrate it more fully into the company. He wanted someone to show the leadership team what it was all about.

Further evidence of how a shift in worldview causes a deep change in leadership approach is given in this highly descriptive (and somewhat indelicate) quotation from a CEO with over four decades of senior leadership experience in two of the largest and best-known American multinational companies in the world:

What I began to experiment with was the change from fear to trust. Trust is a much more powerful tool than fear. Servant leadership is a term that's been over-used now for many years but that's basically where it came from. Also what I saw was a democratization of companies in the mid- to late '90s, away from the typical hierarchical approach that comes from the military, where an officer's idea is better than yours because he outranks you. Good people started to realize that they don't have to put up with that crap.

Another perspective about a shift in leadership is offered by this participant, a senior sustainability consultant who has worked with dozens of major corporate sustainability executives in a consortium. He described his perspective on how sustainability leadership involves the capacity to hold two opposing views at the same time:

I might separate in terms of time-frames. From
an extremely practical standpoint, if we look
at the challenges that individuals are facing and
the situations that they are being put into, first
we have regulatory and control systems. There
is the physical EHS [environmental health and
safety] risk and a reputational risk. Both of these
are about control. On the flip side we have the
opportunity side opening up new more socially
responsible business models. It's tough to find
individuals that can hold both styles. Control
versus opportunity. I see this interesting tension
right now whether they are the CSO, CEO or at
the EVP level, they have to be good at both of
these types of leadership styles.

Another way that the capacity to collaborate more deeply
and widely involved how sustainability executives work with
a wide range of stakeholders including employees, suppliers,
customers, the public sector and NGOs. This participant,
the chief sustainability officer at a global consumer products
company I quoted earlier in Chapter 10, reflected on how
sustainability had changed his way of thinking about
leadership and collaboration in this way:

I was based in Sweden, which has a more
collaborative culture … To get the sustainability
agenda to work internally you have to have
strong influencing skills. When I think about
sustainability leadership in contrast to the 20th
century "great man" style of leadership, you're

more the *collaborator-in-chief*. You're not
command-and-control, you're not omnipotent.

Reflecting further on how his experience with
sustainability has affected his worldview and approaches to
leadership, he went on to add:

> I knew immediately that I did not know everything
> there was to know about sustainability ... I know
> that I will never be an expert. It enforces a new
> level of humility because I know deep down
> that I need to collaborate with others in a more
> flexible way. I'm very grateful that I went into
> sustainability and feel very lucky because it is
> helping me develop some right-brain skills that
> I did not know had gone dormant. Practising
> sustainability has allowed me to collaborate at a
> higher level.

This executive, the senior vice president of global
citizenship at a highly influential global food company, put it
this way:

> Sustainability is the ultimate team sport.

Another sustainability executive added:

> Whenever our team members want to build a
> collaborative effort, it's my job to build that tent.

As part of this theme of collaboration across boundaries,
I found numerous instances in interview transcripts where
sustainability executives shared stories about how their
relationship with the NGO sector had changed. For example,

this senior executive in a global food and beverage company described a large-scale worldwide collaboration with an NGO in the area of water stewardship.

> My focus is on water. We're four years into this
> partnership ... We're focusing on seven major
> river basins and 15 countries. We've melded the
> people in the field so that now they're almost
> indistinguishable ... There is a local connection
> everywhere ... It represents real ignition as to how
> we effectively partner and reach out and engage.

These final three quotations from senior sustainability executives each express how sustainability has led to a deeper collaboration with NGOs in similar ways:

> Collaboration with NGOs at our company has
> changed drastically ... We had to start working
> with and listening to the NGOs at both the local
> and the federal level. You can't please everyone
> all the time but there is a lot of common ground
> on new technologies ... We share lots of common
> purpose with them.

> I would say that one of the biggest shifts and
> mentalities that I have seen in recent years is that
> many people inside our company are now seeing
> sustainability as an opportunity and not just risk
> mitigation, governance or compliance. I think
> what we have been able to do is turn it around
> and engage with society. In the past, engagement
> with society was defined with business being on
> the defensive with society. I think there will be

studies coming out showing how a commitment
to sustainability will be reflected in the corporate
culture and how aligned it is with societal values.

A really good sustainability leader should focus
broadly on a sustainability strategy and make it
come alive. We've really focused on engagement
with stakeholders. Sometimes it takes a lot more
listening. We created a human rights policy based
on feedback with stakeholders in key issues. It has
been dramatically important to the point now we
have close relationships with numerous NGOs and
advocacy groups. They now know that we will do
the right thing with the right feedback. So now we
find out about things early in the curve.

Another highly collaborative approach to leadership that
appeared in the interviews was an enhanced capacity to
translate and communicate complex sustainability initiatives.
This capacity arises from the greater awareness of diverse
worldviews and contexts that is a key characteristic of a
post-conventional worldview. Participants spoke of their
challenges to communicate to diverse stakeholders and
create shared value. For example, this executive involved in
global food issues at a large environmental NGO shared this
perspective:

Last week at a panel I talked about how the
shared value frame allows companies to say
that they have the technical fixes to social and
environmental problems as well as the resources
and expertise to deliver them. And that we can

now make the business case that it's going to
provide us with cost savings or it's going to give
us reputational improvement or what have you.
However, it still is coming down to that, although
we have many of the technical solutions faced by
people around the world, there are political and
economic forces in play that make this too little
too late. There just aren't win–win scenarios to
be had everywhere, especially on political issues.
The power dynamics and economic inequalities
are uncomfortable. There is not really a vernacular
for the overall context. We've got to find some
language that would be translatable for companies
and governments.

This executive, the chief environmental strategist
at a global technology firm, described his view on
communication this way:

Most companies are in a state of advanced chaos.
The problem is communicating to everyone
around the world. The reality today is that there
are so many channels out there, we have to get
the word out that we are doing this at a level
that we've never heard of before. The key is to
have voices go viral and rise above the clutter.
My challenges are more about getting people to
understand what is happening and change their
behaviour at scale.

This long-time senior sustainability executive at a major
apparel and footwear company I quoted earlier shared her
observation about communication this way:

> What we've found in regards to our sustainability
> initiatives is that people go through phases of
> understanding over a long period of years. This
> means that we have to continually evolve our
> communication.

Yet another example of the capacity to communicate
to diverse worldviews can be found in the words of this
president of a private company included earlier in Chapter
10:

> Part of the reason is that I don't stand on
> a soapbox is to allow a constant drip of
> improvements that starts to take root, one by
> one. I don't ask employees as to how they regard
> *our* efforts. I want my conversations with my
> team to be about *them*, and *their* needs ... It's my
> experience that I regard these issues different than
> most people, I don't want to environmentalize as
> a religion. I'm motivated on a very practical level.
> There's a risk if I were to be seen as too extreme in
> my views. It's kind of a circuitous pattern.

The idea of translation also arose from this executive while
she was framing the challenges of communicating the need
to reduce carbon emissions.

> It's not going to be easy to translate to the
> common consumer that it is the right thing to do
> to reduce carbon emissions.

Finally, this senior sustainability consultant, who has
worked extensively with corporate sustainability executives

and facilitates a large support network, provided this creative new job title that captures the way sustainability executives approach their work:

> We all know that we have to have a steady drumbeat to make a difference. I wrote something a while ago that said the CSO should be called the "Chief Translation Officer".

The quotations above demonstrate new ways that these executives are changing the way they think about sustainability leadership. We can see in their words how their worldviews appear to be enabling their capacity for collaboration on a deeper level. The capacity for deep collaboration also leads to breakthrough innovation, something Harvard management professor Linda Hill calls collective genius.[94]

> The post-conventional archetype of "collaborator-in-chief" can lead to new breakthrough environmental technologies

When fuelled by an ecological worldview, the post-conventional archetype of "collaborator-in-chief" can lead to new breakthrough environmental technologies and suggests how sustainability leadership must evolve in the future. In the next chapter, I use the cumulative insights from the research to make specific recommendations to cultivate a new psychology for sustainability leadership.

12
Cultivating a new psychology for sustainability leadership

At the beginning of this book, I reflect on the limits of the term "sustainability" and how it can mean many vastly different things to different people, from business as usual short-term economics to new long-term deeply ecological and restorative business models. As Professor Bob Brinkmann, Director of Sustainability Studies at Hofstra University observes, "Sustainability is in the eye of the beholder."[95]

What has been missing is a way to reclaim the term, and the movement, in a way that captures the latter meaning. In this spirit, I propose that as sustainability educators, consultants and executives, we need to develop a new type of shared language. This can

> As sustainability educators, consultants, and executives, we need to develop a new type of shared language

be accomplished through a new psychology for sustainability leadership.

In the preceding chapters, I have taken steps in this direction. The goal for my research was to understand better the interior worldviews of global sustainability leaders. In order to explore their worldviews from diverse perspectives, I drew on several social science disciplines that up until now have for the most part not been used to study sustainability leadership.

By using insights from these disciplines to guide the analysis of the interviews, evidence was uncovered that led to four major themes. In Chapter 4, we explored **experiences that shape ecological worldviews**. In Chapter 7, I presented **expressions of ecocentric worldviews and ecological self**. In Chapter 10, we explored **expressions of post-conventional worldviews**. Finally in Chapter 11, I described their approaches to sustainability leadership through the theme of **collaborator-in-chief**.

The four major themes and supporting findings offer new insights into the worldviews and motivations of individuals behind sustainability initiatives in multinational corporations. These insights are significant in five ways:

1. The descriptions of how global sustainability leaders think about nature, and where these thoughts came from, indicate that they have developed *explicit* ecological worldviews from specific sources of origin.

2. Sustainability leaders appear to make a connection between how they think about the natural world and the motivation for their work in sustainability.

3. Their ecological worldviews appear to have been formed throughout their life-span.

4. The findings suggest that many sustainability leaders possess a high degree of ecological intelligence on a planetary scale and also have a philosophical stance on their relationship to nature. Many of them appear to be not only highly educated in the complexity of global environmental science, but also readily aware of bigger philosophical questions facing humanity in regard to our relationship to nature.

5. The interviews contain ample evidence of post-conventional worldviews that reveal how they think about their role as sustainability change agents and how this influences their approaches to leadership, collaboration and communication.

Each of these themes and supporting findings invites new areas of sustainability leadership-related education and research. Here are several specific recommendations for business educators, sustainability executives, consultants and activists to consider:

12.1 Integrate eco-social sciences in the business curriculum

In order to cultivate ecological worldviews in students and the next generation of sustainability-minded executives, a full range of eco-social science literature and research needs to become part of the mainstream business curriculum. Currently, exposure to eco-social sciences in business education is limited to a select group of sustainability-specific programmes, and mostly at the graduate level. The eight perspectives on ecological worldviews and the specific scholars cited in Chapter 3 offer an initial roadmap for this purpose.

For example, in a standard semester, an overview of each of these eco-social science disciplines could be presented to students through select readings and weekly reflective papers. This type of interdisciplinary curriculum can be offered in traditional face to face, online and hybrid class formats.

Key themes could include:

- Implications of anthropocentric worldviews

- Shallow versus deep ecology

- Human relationship with technology

- A new ecological paradigm

- The ecological self

- Integral ecology

- Valuing natural capital

- Social psychological contexts for climate change

- Indigenous worldview

Through weekly discussion forums, students can reflect on how each of the social science disciplines and its respective themes contributes to an ecological worldview. Through these types of reading and discussion, a fuller understanding of ecological worldviews and their implications for decision-making can stabilize in the minds of students and young executives.

Among the many dozens of references provided at the end of this book that serve this purpose is *Eco-Literate: How Educators are Cultivating Emotional, Social, and Ecological Intelligence.* In this recently published "how-to guide" for educators, Daniel Goleman and his co-authors Lisa Bennett and Zenobia Barlow explore the applicability of emotional, social, and ecological intelligence to education.[96]

They suggest five practices for emotionally and socially engaged eco-literacy:

1. Developing empathy for all forms of life

2. Embracing sustainability as a community practice

3. Making the invisible visible

4. Anticipating unintended consequences

5. Understanding how nature sustains life

Although not explicitly framed in the context of business and sustainability leadership, each of these practices supports the development of ecological and post-conventional worldviews for the next generation of business leaders.

12.2 Reflective journaling and eco-biography exercises

Another important new element in the business curriculum would allow students to explore the significant life experiences that have shaped their worldview resulting in a more explicit "eco-biography". By becoming familiar with each of the five significant life experiences presented in Chapter 4, students and executives can generate powerful new insights about how they became who they are and uncover new connections between their thoughts about the natural world and their motivation for their work in the sustainability field.

The nature writer John Daniel describes the benefits of understanding the story you are in, based on an understanding of your experiences early in life.[97] He observes that understanding how those experiences have shaped who you are can yield a greater self-awareness and a more vigorous self-reliance. A more vigorous self-reliance based on a new level of eco-psychological self-awareness can potentially enhance our capacity as sustainability change agents to solve

> Understanding how those experiences have shaped who you are can yield a greater self-awareness and a more vigorous self-reliance

wicked problems, maintain resilience and overcome challenges.

What distinguishes the individual corporate sustainability leaders I interviewed for this book in the general population of corporate executives is that their life experiences appear to have contributed to the formation of an advanced ecological worldview.

You will recall from Chapter 4 that the five findings also appeared to suggest a chronological sequence. The first significant life experience about early childhood generally referred to the K-12 years. The second that focused on environmental education generally corresponded to their college and graduate school years. The third life experience about witnessing extreme poverty and environmental degradation for the first time in developing countries appears to have occurred a bit later in their 20s. The fourth that relates to career shifts where they became aware of the impact of global corporations on planetary ecosystems and social issues appears to have taken place primarily when they were in their 30s. Lastly, the fifth significant shift where they describe spirituality and a sense of service generally appears to be a reflection of their current lives, which would correspond to their late 30s or 40s for the large majority of the participants.

Additional significant life experiences that contribute to an ecological worldview would be uncovered through these types of reflective exercise by students and executives. Over time, these types of eco-biography could become a standard eco-psychological foundation for sustainability leaders.

12.3 New developmental assessments for ecological worldviews

The five significant life experiences also suggest a possible developmental sequence. This could potentially be used to create a new type of developmental scaffolding for sustainability leadership through further research by eco-social science, developmental psychology and organizational leadership researchers.

However, this inference is limited by the qualitative exploratory methodology and the specific interview questions I used in this study. As a result, the suggestion that these experiences represent a hierarchical and developmental sequence is tentative and would need to be supported by further empirical research. This could be approached through additional semi-structured interviews focused on a developmental line of inquiry.

Another approach could be to modify existing leadership assessment tools to explore a stage conception for ecological worldviews. Integral ecologists Sean Esbjörn-Hargens and Michael Zimmerman offer a model of ecological selves and worldviews, but thus far in an integral theory-building mode. Many studies have been conducted using the new ecological paradigm (NEP) survey instrument, but they have focused primarily on environmental values, education and public policy, not on sustainability leadership development. In addition, the NEP has been criticized from a psychological perspective as I describe in Chapter 3.

Through their decades of extensive empirical research, developmental theorists have accumulated an immense body of work that has established the validity of developmental theory and focused on a wide range of applications. However, for the most part their research has taken place under the anthropocentric umbrella of Western psychology. Thus, it has focused primarily on the relationship of human beings to themselves and to each other, and not on the relationship to nature.

Among the primary psychometric instruments used for their research include the various versions of the Washington University Sentence Completion Test, such as the Global Leadership Profile and the Leadership Maturity Framework. The significant life experiences presented in Chapter 4 could be used to integrate ecological worldviews into their theories and measurement instruments. This could be accomplished by adding eco-sentence stems to explore a stage conception for ecological worldviews, thereby increasing its relevance for sustainability leadership development.

12.4 Corporate workshops to support the ecological self

The interview narratives presented in Chapter 7 demonstrate how several participants appear to have an awareness of their embeddedness in the natural world, one of the core aspects of the ecological self. This serves to ground the construct of ecological self further in sustainability practice and animates key insights from the fields of eco-psychology, integral ecology and deep ecology. It offers a new link between the ecological self and deeper motivation for sustainability leadership.

This may allow new ways for sustainability leaders to understand themselves and ultimately enhance their determination and effectiveness as change agents. Most importantly, it suggests the ecological self could become a new type of interior psychological foundation for sustainability leadership development. This could be accomplished by integrating existing research on the ecological self in sustainability leadership education and corporate training programmes.

> The ecological self could become a new type of interior psychological foundation for sustainability leadership development

As described earlier, human development research from eco-psychology, deep ecology and integral ecology suggests that the ecological self is part of an expanded self-concept that can significantly change how an individual acts in the world. This research further suggests that as human beings we may be underachieving our self-potential by not embodying our ecological self.

The deep ecologist Bill Devall observed that as human beings we identify ourselves as our religion, our gender and our occupation, and that we underestimate our self-potential by not identifying with our ecological self. It's interesting to note that the occupation of the sustainability executive did not yet exist until very recently. Devall invites us to consider that the ecological self is not static but a search for an opening to nature.[98]

Although this gradual opening to nature may be one of the most important psychological drivers of sustainability leadership, we have yet to develop a specific developmental model for how this can actually happen over the course of one's life and work in the sustainability field.

Although the transformational workshop created and facilitated by Joanna Macy, John Seed, Molly Brown and many others called **The Council of All Beings** and the **Work That Reconnects**[99] has been spread widely over the last several decades, their work has not received enough attention within the corporate world or in academia. New workshop initiatives could also include new adaptations of Sewall's **five perceptual practices**[100] and Esbjörn-Hargens and Zimmerman's integral model of **ecological selves**.[101]

There is a fascinating question for us to consider. Does the practice of sustainability catalyse the development of the ecological self? Or, do individuals with a sense of their ecological selves select work in sustainability? Either way, a new opening, or deepening of the ecological self, may be occurring through the practice of sustainability. This highlights the potential for the ecological self as part of sustainability leadership development in the years ahead.

> A new opening, or deepening of the ecological self, may be occurring through the practice of sustainability

12.5 Greater developmental focus on corporate sustainability leaders

Among the major themes and findings from the research, the most surprising was the ample evidence of post-conventional worldviews in the sample of global sustainability leaders. The post-conventional capacities include an expanded awareness of context and worldviews, thinking in time-horizons longer than decades, holding an enhanced systems consciousness, having expanded circles of identity and care, and a consistent capacity for inquiry.

While it may be expected to find a large percentage of highly developed ecological worldviews among global sustainability leaders, the evidence that a large percentage may also possess the capacities that are consistent with post-conventional worldviews presents an important new line of inquiry based on developmental psychology in the context of sustainability leadership.

The ample evidence of post-conventional worldviews in the study sample holds numerous implications for the field of developmental. Based on several large-scale studies, developmental researchers have found the overall distribution of professional adults in the United States with post-conventional worldviews to be approximately 6–7%. The relatively small percentage of adult professionals with post-conventional worldviews was found to be virtually identical using Torbert's profile instrument (total $N = 497$) and Kegan's subject–object interview (total $N = 342$).[102]

Although there are methodological limitations to this study that are discussed in Appendix B, the qualitative evidence suggests that the sustainability leaders included in this study exhibit post-conventional worldviews at a higher percentage than the general population of adult professional and non-sustainability-oriented corporate executives. When considering that the participants in this study operate inside some of the world's most influential companies, the potential benefits of following this line of inquiry is enormous.

12.6 Integrate developmental psychology in the sustainability curriculum

The evidence that the interviewees' worldviews appear to be manifesting in several highly collaborative approaches to leadership demonstrates the hidden power of using worldview research to advance sustainability leadership. These new leadership approaches include an enhanced capacity to communicate complex sustainability initiatives to a diversity of worldviews, influencing without control, and using an enhanced systems consciousness to work more effectively on social and environmental issues with a wide range of external stakeholders including NGOs and government.

> It became apparent that the practice of sustainability was driving an evolution in their thinking about leadership, collaboration and change

In response to questions about how sustainability was being integrated into the leadership and culture of their companies, it became apparent that the practice of sustainability was driving an evolution in their thinking about leadership, collaboration and change. This represents a new way that the existing body of organizational leadership theory that integrates developmental theory can contribute to the psychology of sustainability leadership.

Among the most prominent leadership books and theories that use developmental psychology to understand organizational leadership are *Action Inquiry: The Secret of Timely and Transforming Leadership* by Bill Torbert and associates, *The Practice of Adaptive Leadership* by Ronald Heifetz, Alexander Grashow and Marty Linsky,[103] *Leadership Agility* by Bill Joiner and Stephen Josephs,[104] and *Immunity to Change* by Robert Kegan and Lisa Lahey.[105] However, this pioneering canon of work has for the most part not placed its primary focus on sustainability leadership development. As a result, it has not received enough attention from sustainability educators.

In their pioneering article entitled "Seven transformations of leadership", voted one of the top ten best leadership articles in the *Harvard Business Review*, David Rooke and Bill Torbert explain how leaders can progressively move through seven stages of development. Based on research with thousands of executives in diverse industries over 25 years, they have found that leaders with post-conventional action logics/worldviews are able to consistently innovate and transform their organizations. They describe the specific capacities that these leaders use to lead successful organizational transformations and generate superior results. These capacities include greater communication and collaboration skills, enhanced systems thinking, and operating with a more strategic time-horizon.[106]

Placing their research in the context of sustainability leadership, Rooke and Torbert cite Joan Bavaria as an example of a leader with a post-conventional worldview. Bavaria is the CEO of Trillium Asset Management, the first socially responsible investment fund. She was also one of the co-authors of the CERES Environmental Principles which, in co-ordination with the United Nations, led to the Global Reporting Initiative, one of the most important sustainability milestones in history. Through the example of Joan Bavaria, they make the link between post-conventional worldviews and the psychology of sustainability leadership.

> Through the example of Joan Bavaria, they make the link between post-conventional worldviews and the psychology of sustainability leadership

12.7 Narrowing the gap between thought and action

In a recent study conducted by MIT and the Boston Consulting Group entitled *Sustainability's Next Frontier*, researchers explored the extent to which corporations are addressing sustainability issues. Based on a worldwide sample of corporate leaders, they found that although nearly two-thirds rate social and environmental issues as significant, less than 10% report that their corporations are addressing them thoroughly. The researchers conclude by attributing this gap to a "disconnect between thought and action".[107]

Beginning with C.S. Peirce in the late 19th century, pragmatist philosophers including William James and John Dewey put forth the notion that the deepest meaning of a proposition is found in the practical consequences of accepting it. The central proposition of this book is that the extent to which we think ecologically ultimately drives the depth and effectiveness of our action towards sustainability. As we near the end of this book, I invite you to consider and reflect on the practical consequences of accepting this central proposition.

If, as Mahatma Gandhi reminds us, our thoughts become our actions, it follows that how we think about our relationship to the natural world drives our action towards it. In other words, a better understanding of ecological worldviews leads to a new understanding of *why* and *how* we act as sustainability leaders. In terms of impact, our worldviews powerfully shape the specific choices we make and drive *what* sustainability initiatives will be accomplished. Ultimately, this means the extent to which a corporation may accomplish decarbonization, water stewardship and a host of ecologically restorative business practices on a global scale. Figure 12.1 illustrates this central idea.

> A better understanding of ecological worldviews leads to a new understanding of *why* and *how* we act as sustainability leaders

FIGURE 12.1   The hidden power of ecological worldviews

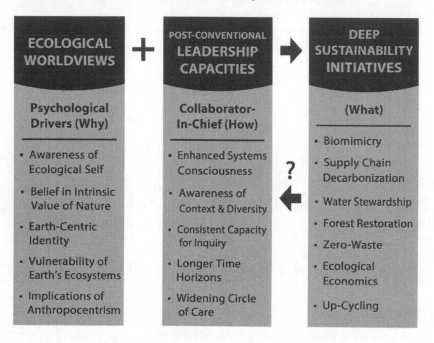

By providing evidence of the ecological worldviews of senior sustainability executives and how this relates to their motivation, this research starts to close this gap between thought and action for sustainability in global organizations. When combined with how post-conventional worldviews affect their approaches to leadership and collaboration, we can begin to appreciate fully the hidden power of ecological worldviews.

12.8 Future research questions

Through its interdisciplinary framework, this book
sets the stage for eco-social science researchers from
numerous disciplines to pursue a major new area of focus:
sustainability leadership. Although a few studies have been
conducted in this area, an understanding of how ecological
and post-conventional worldviews can advance the field of
sustainability leadership is just beginning.

Future studies with new samples could explore such
questions as:

- How can an understanding of ecological and
  post-conventional worldviews be integrated into
  sustainability leadership development programmes?

- What types of new corporate workshop and
  intervention can accelerate the development of the
  ecological self?

- How can a more comprehensive understanding
  of worldviews from the social sciences be used to
  enhance sustainability-related communication?

- How can we develop new assessment instruments
  designed to measure ecological worldviews in
  sustainability leaders?

- How do ecological worldviews vary across age
  groups, gender and culture?

- How are ecological worldviews expressed at specific
  developmental stages?

- How do the ecological worldviews of individual leaders correlate with the success of specific sustainability initiatives?

While the exploratory nature of this book does not fully provide answers to the above questions, it offers initial steps towards this undertaking and cultivating a new psychology for sustainability leadership.

13
Multinational executives as human trim tabs

A "trim tab" refers to the tiny piece at the edge of a rudder that is used to turn an enormous ship. The philosopher, architect and systems theorist Buckminster Fuller was the first person to use the term when referring to human beings. Since then, systems thinkers have used the term "trim tab" to explain how the actions of just one person can ripple out into our global society and cause large-scale social and environmental change.[108] Fuller also coined the term "Spaceship Earth" and when he died he had the epitaph "CALL ME TRIM TAB" engraved on his headstone.

> Systems thinkers use the term "trim tab" to explain how the actions of just one person can ripple out into our global society and cause large-scale social and environmental change

Pioneering consciousness researcher Jean Houston has used the term "trim tab" to describe the potential for leaders to transform society as part of her work in social artistry.[109] The cultural anthropologist Margaret Mead reminds us to "never doubt that a small group of thoughtful committed citizens can change the world; indeed, it's the only thing that ever has". Although these words historically been have been used to inspire social and environmental movements, as we come to the end of this book I propose we apply them to senior sustainability executives in multinational corporations.

Among the many corporate/NGO collaborations in recent years, one that I follow closely is called BICEP, which stands for Business for Innovative Climate and Energy Policy.[110] I first became aware of BICEP several years ago when I spoke with executives from several participating companies as I was beginning the research that led to this book. What I remember thinking at the time was that these individuals represented a new type of executive that ultimately could have increasingly positive impacts on the many challenges facing our environment and human communities around the world.

At the time, I presented a paper at a sustainability conference entitled: "The emergence of the chief sustainability officer: A new archetype in corporate leadership". I explained how inside multinational corporations are a growing number of highly influential and environmentally aware corporate executives that collectively represent a new breed of business leaders. I described how they came from diverse backgrounds from engineering

to environmental science and how many possessed
a sophisticated understanding of natural resources,
climate change science and the socioeconomic realities
of communities around the world where their companies
operate.

I observed that these executives not only had to be capable
of navigating the short-term financial realities of quarterly
earnings, return on investment and market share, but they
also balanced these demands with increasingly complex
environmental and social challenges as part of their job
description. However, what I was unable to articulate in
my mind at that time was that I was searching for a better
understanding of their worldview.

Shortly after the pioneering sustainability CEO Ray
Anderson passed away in 2011, GreenBiz Executive Editor
Joel Makower asked why there have not been more CEOs
with the same depth of understanding and commitment
to sustainability as Anderson.[111] Recently, after the 2014
Net Impact Conference,[112] green business journalist Mark
Gunther asked a similar question, only he acknowledged
Unilever CEO Paul Polman as perhaps the only current
leader of a major public company thoroughly dedicated
to principles of sustainability. In the same article about
Starbuck CEO Howard Schultz, Gunther went on to ask a
bigger question: Why aren't there more CEOs willing to put
society's social and environmental needs at the core of their
business?[113]

The research presented in this book presents a new type
of answer to this question. It suggests that many of the
most influential sustainability executives are motivated

by their ecological worldviews. We know from the field of organizational development that senior leaders surround themselves with a team operating from similar worldviews. In other words, CEOs with an ecological worldview will not only be making decisions based on this perspective, they will surround themselves with dozens of senior executives with ecological worldviews. Perhaps we need to find new ways to develop future CEOs with the ecological worldviews of today's sustainability executives?

> CEOs with an ecological worldview will not only be making decisions based on this perspective, they will surround themselves with dozens of senior executives with ecological worldviews

Human beings now face the most serious and complex set of ecological problems in our history. Multinational corporations must play an important role in solving the planet's great ecological challenges. Taking the courageous and transformative actions to redesign our global economic system to bring it into alignment with the Earth's ecosystems will ultimately need to be driven by a large-scale shift in worldviews by a critical mass of global corporate leaders.

Collectively, multinational corporate executives are the most powerful human force on the planet. And perhaps the most powerful force driving the actions of these executives, and ultimately the actions of their global organizations, will be their worldviews.

Appendix A: Ecological Sustainability Worldview Assessment Tool (E-SWAT)

My intention here is to help you apply the new concepts presented in this book to your development within your context for sustainability. By becoming familiar with eco-social sciences, development psychology and their key concepts, each of us can experiment with where we gain the most traction and insight.

> Each of us can experiment with where we gain the most traction and insight

Through a new self-assessment called the E-SWAT and reflective exercises, this appendix offers you new ways to explore how ecological and post-conventional worldviews can help you understand yourself better as a sustainability leader. It encourages personal reflection on the deeper motivations for your work in sustainability, how you think about nature, and how this can deepen your approach to sustainability leadership. This in turn can help you expand the ways you communicate, improve your resilience in the face of challenges, and ultimately enhance your long-term effectiveness as a sustainability change agent.

As Francis Moore Lappe tells us in *EcoMind*, "Seeing determines our capacity for doing … Our ability to solve a problem is affected by how we how we frame it in our minds and how we perceive the challenge … We each can reframe our thinking in ways that give us energy to engage."[114]

Since I first started teaching leadership at both the undergraduate and graduate levels ten years ago, I've been meditating with my students at the beginning of class. Always starting with just a few simple breaths, I've found this helps students relax, reduce their stress and let go of at least a few things on their minds. In this way, they can become a little more present for whatever we are going to experience together in class. For the majority of students, I'd say on average more than 90%, this is their first experience with meditation, and surprisingly this percentage has not changed over the last ten years in my observation. As university students and executives, cultivating peace of mind or stress reduction through meditation is still unfortunately not the norm in our busy lives.

> Since I first started teaching leadership at both the undergraduate and graduate levels ten years ago, I've been meditating with my students at the beginning of class

However, across virtually all spiritual traditions, meditation, contemplation and prayer are the most reliable ways to cultivate a steady mind and gain insight into whatever area of our lives is needed. It follows then, that in order to gain insight into our worldviews and our motivations for sustainability, meditation practice can help.

My influences range from Buddhist practice, Western depth psychology and indigenous nature-based practices. It also includes the Hero's Journey made famous by Joseph Campbell[115] and men's shadow work as part of the ManKind Project.[116]

However, with students at the beginning of each class, I try to keep it simple. I offer just a few instructions to relax and sit comfortably, become aware of the breath moving in and out, the sounds in the room and outside, sensations in the body, the thoughts in the mind coming and going. The reason I start with meditation at the beginning of this chapter is the same reason I start with it at the beginning of class ... so that I can now invite you to take a few breaths and try to become a bit more aware of your current state of mind before exploring the following exercise on worldviews. Try it now.

> Become aware of the breath moving in and out, the sounds in the room and outside, sensations in the body, the thoughts in the mind coming and going

Another reflective exercise I like to do with sustainability leadership students involves personal visioning or mastery. I first read about personal mastery in *The Fifth Discipline* by Peter Senge, who credits teacher/composer Robert Frtiz for originally exploring the practice.[117] It involves the ability to see your current reality, then articulate your personal vision around sustainability, and finally make a commitment to achieve your desired results while working with the creative tension between the two. It's based on extensive research that suggests that, for deep and enduring learning to take place, it needs to be related to a person's own vision and personal interests.

> Make a commitment to achieve your desired results while working with the creative tension between your current reality and your personal vision

The challenge is not to let practical challenges limit your vision. As Senge describes, this process activates a subconscious level of mental activity that creates sustained energy for personal growth.

So as you begin to work with the concepts from this book, imagine how they can lead to various results in your sustainability career that you deeply desire. Ignore how possible or impossible your vision seems. Allow yourself to think long-term, perhaps as many as ten years out into the future.

The E-SWAT

One of the most common tools used for strategy and business planning is called the SWOT, which stands for strengths, weaknesses, opportunities and threats. I'd like to introduce you to the E-SWAT, which stands for Ecological Sustainability Worldview Assessment Tool. The E-SWAT contains eight separate steps and is based on the major themes that are presented in this book. It is designed to help you establish a baseline understanding of your sustainability worldview through the numerous theoretical lenses introduced in the preceding pages. By comparing your reflections from the E-SWAT with those of other people, it may help you explore further how your worldview may align or differ from others with whom you work, and open new ways to communicate and influence change.

> The E-SWAT may help you explore further how your worldview may align or differ from others with whom you work, and open new ways to communicate and influence change

Step 1

Read through the following list of items and make a check mark next to each item that has influenced the development of your ecological worldview over the course of your life.

- Growing up in nature

- My immediate family

- K-12 outdoor education

- College or graduate school environmental education

- Seeing poverty or environmental degradation while travelling abroad

- Hunting, camping, fishing or other experiences in nature as an adult
- Perceiving capitalism as a vehicle for social and environmental change
- A sense of service
- Spirituality
- Other _____

Step 2

Next, go back through each item and make a few notes about your specific memories of those experiences and how they may have contributed to your ecological worldview and how it has continued to evolve over the course of your life.

Step 3

Now, considering your new awareness of where your ecological worldview comes from, what experiences would you like to have in the future that can deepen or enhance your worldview? Make a list. This can be a book you're interested in reading, a group you are interested in joining, a class or seminar, or an educational trip you are interested in taking in the future.

Step 4

Next, read through the following list of items and make a check mark next to each item that forms part of the motivation for your work in sustainability.

- My awareness of an ecological crisis
- My belief in the limits to economic growth
- My belief in the intrinsic value of nature
- My belief in the role of technology to solve ecological challenges
- My belief in capitalism as a vehicle for social change
- My belief in capitalism as a vehicle for environmental change
- My religion
- A sense of service
- My spirituality
- Other _____

Step 5

Considering your answers thus far, explore how your ecological worldview relates to your motivation for your work in sustainability. Where do you see connections? Where do you gain the most insight? What do you want to learn more about?

Step 6

Read through the following list of items and rate the extent to which you agree or disagree with the following statements: (a) not at all, (b) occasionally or (c) frequently.

- I am aware of my own worldview
- I am aware of the worldviews of others
- I am aware of the context for my work in sustainability
- I am aware of the context of others with whom I work
- I am aware of the many systems that I am part of in my work
- I think about sustainability in a historical and long-term future context
- I think of myself deep down inside as a global citizen
- I think of myself deep down inside as one of many species on the Earth
- I consistently think about my self as part of larger systems
- I consistently maintain a capacity for inquiry, even while under stress

Step 7

Based on your initial answers to the above items and what you have read in this book, describe any additional thoughts or questions you may have about developing post-conventional worldviews and how they may relate to your work in sustainability.

Step 8

Lastly, discuss what you have learned about yourself through the above steps with a fellow student, colleague at work, friend or family member as a way to deepen your journey into the hidden power of worldviews and cultivate your new psychology for sustainability leadership.

Appendix B: Research methodology and description of participants

In the design of my research, I used a qualitative exploratory methodology.[118] Utilizing the ten-question interview guide shown below, I conducted interviews with 75 sustainability leaders using principles of naturalistic inquiry.[119]

Although I met many of the participants face to face at conferences or at their corporate headquarters, most of the formal interviews took place over the phone. The interviews typically lasted from 30 to 45 minutes and were transcribed by me either during or immediately after the interviews took place.

Recruitment of participants and description of sample

I used a purposive sampling strategy focused on senior sustainability executives at multinational companies and recruited participants primarily by attending a dozen different national and international corporate sustainability conferences over three years. As I began conducting interviews, I also deployed a snowball sampling strategy as sustainability executives referred their colleagues at other corporations or NGOs, who then became additional participants in my study.[120]

Of the 75 participants that comprised the sample, 54 held senior-level positions in multinational companies at either the chief sustainability officer, vice president, director or manager level. There were three CEOs of public companies, six presidents of private companies, six senior executives of environmental NGOs, and six sustainability consultants.

The companies represented in the sample were primarily involved in consumer goods such as food and beverages household products and consumer electronics. A partial list of the companies included Hewlett-Packard, Microsoft, Aveda, Clorox, MillerCoors, Sprint, AT&T, Motorola, AMD, Waste Management, 3M, Mattel, Starbucks, Nike, SC Johnson, Seventh Generation, Coca-Cola, Ford, GE, Price Waterhouse Coopers, Sun Microsystems, Green Mountain Coffee, Albertsons and Ben & Jerry's, which is a subsidiary of Unilever.

The range of job titles included Chief Sustainability Officer, Vice President of Global Citizenship, Vice President of Environment and Water, Director of Sustainability and Stewardship, Director of Natural Resources, Director of Social Mission and Manager of Product Integrity. A few of the most interesting titles were Director of Stakeholder Mobilization, Director of Coffee Community Outreach and Director of Corporate Consciousness.

All the participants had at least five years of experience co-ordinating and communicating sustainability-related initiatives to a broad range of internal and external stakeholders including their employees, supply chains, NGO partners and customers. Many of the participants had worked closely with corporate sustainability initiatives for more than ten years and had held multiple senior positions in more than one multinational corporation. Many had worked in both the private and public sector and several had made the move from environmental NGOs to executive positions with multinational corporations.

Analysis and interpretation of the qualitative interview data

In order to analyse and interpret the qualitative data that I collected through my semi-structured interviews, I utilized a multistep thematic analysis process.[121]

First, an inductive thematic analysis process was conducted in order to get a general sense of the information being conveyed and uncover initial themes from the interviews. This involved reading over the interview transcripts multiple times, coding key themes that were supported by diverse quotations from the participants. After several rounds of coding and analysis I identified the five significant life experiences that shape ecological worldviews presented in Chapter 4.

Second, a deductive thematic analysis process was utilized to analyse the qualitative interview data based on the descriptions of ecocentric worldviews and the ecological self from my review of the literature. After several rounds of coding and analysis I identified the five expressions of ecocentrism and ecological self that I presented in Chapter 7.

Third, I deployed a deductive thematic and hermeneutic methodology to analyse the interview transcripts based on the descriptions of post-conventional worldviews from my review of the literature. Hermeneutics can be defined as the art and science of interpretation of texts.[122] As part of qualitative research, hermeneutic methodology describes a type of multilevel interpretive process that researchers can use to further analyse interview transcripts through a series of hermeneutic turns.[123] This involved becoming more aware of the context of each participant at the time of our interview, my intention and context at the time, and the current context within which the data was being analysed. It also led to an awareness of the context of the theories that were being used to analyse the interview texts.

After more than a dozen rounds of coding, sorting and hermeneutic analysis, I identified specific groups of key words and phrases used by the participants that led to the five distinct ways that sustainability leaders express post-conventional worldviews presented in Chapter 10.

Lastly, after further rounds of inductive thematic analysis, I identified the final set of findings described as the "collaborator-in-chief" in Chapter 11.

Limitations of the sample

The executives I interviewed came primarily from a subset of companies in consumer goods including food and beverages, household products and consumer electronics. Prior to my interviews, in most instances I had over several years—with my students as part of my "Case studies in corporate sustainability" classes—studied their balance sheets, income statements and sustainability reports. As a result, the companies were selected based on the external reporting of the ways they were reducing ecological footprints and engaging in a wide range of progressive social initiatives around sustainability. Although I did not apply specific quantitative criteria for their selection, they clearly represented a very biased sample.

Corporations directly involved in certain industries such as petroleum extraction, natural gas fracking, defence contractors and genetically modified food were not included in the sample. Therefore, the findings made in this study are based on data gathered from sustainability executives only within certain industries and therefore limit their generalizability to these industries.

Limitations of the methodology

The expressions of post-conventional worldviews were identified through a deductive thematic analysis of textual data and a hermeneutic methodology based on my biased sample. Participants were not assessed with a psychometric instrument such as the Global Leadership Profile (GLP) or the Leadership Maturity Assessment (MAP). I also did not interview executives from a control group of "non-sustainability executives". As a result, the suggestion that the sustainability leaders included in my study sample exhibit post-conventional worldviews at a higher percentage than the general population of business leaders is only tentative.

However, through the analysis of textual data, it does strongly suggest a higher percentage of post-conventional worldviews exists in the sample population of global sustainability leaders than exists in the general population of corporate leaders. In order to further validate this finding, additional empirical data would need to be gathered from a control group by means of a psychometric instrument such as the GLP or MAP.

This study also does not make claims as to the specific stage of development of the participants or the precise percentage of participants that would have been assessed at each stage, or make a distinction between the three specific levels of post-conventional worldviews. Based on data gathered through validated psychometric instruments, developmental researchers have found that leaders can progressively move through seven progressively more complex stages of psychological development. What they call the "transforming stage" is the middle of the three most advanced stages of development they call "redefining", "transforming" and "alchemical". For the purposes of my research, I included these three advanced stages in the broader stage of development known as a post-conventional worldview.

Lastly, I did not use statistical analysis or make an attempt to correlate the individual executives I interviewed with specific sustainability initiatives and overall successes of their companies. Given the number of variables involved including the influence of their CEOs, boards of directors, changing economic conditions and a wide range of stakeholders, this was beyond the scope of my research design. However, it is my hope that this book will lead to new research by social science scholars with new samples of sustainability leaders in the years to come.

Semi-structured qualitative interview guide

1. Perhaps we can start with some general background. How, or why, did you become involved with sustainability within your organization?

2. Where do you think your deeper motivation comes from in regard to these issues?

3. How would you describe your worldview as relates to nature and ecology? What comes up for you when you think about your relationship with nature?

4. Looking back, can you point to any transitions or events where you started to look differently at the world, yourself and nature, or is this a worldview that you have held for a long time?

5. How do you think that your work in sustainability has had an impact on your worldview?

6. How do you perceive global environmental issues today and what you see as the source of many of the problems?

7. Can you think of a situation or a dilemma where your worldview was in conflict with an activity you were involved in as part of your work? How did you resolve this?

8. What do you believe are some of the implications of ecological worldviews on sustainability leadership development in general?

9. What do you see as your biggest challenges to accomplishing your goals at both the individual and organizational levels?

10. Is there anything that we did not touch on or that you would like to share before we wrap up?

Notes

Preface

1. E. Weinreb (2011) *CSO backstory: How chief sustainability officers reached the C-suite*. Retrieved from http://weinrebgroup.com/wp-content/uploads/2011/09/CSO-Back-Story-by-Weinreb-Group.pdf.
2. L. Brown (1995) Ecopsychology and the environmental revolution: An environmental foreword. In T. Roszak, M. Gomes & A. Kanner (Eds.), *Ecopsychology: Restoring the Earth, healing the mind* (pp. xiii-xvi). Berkeley, CA: Sierra Club Books. p. xvi.
3. B.C. Brown (2012) Conscious leadership for sustainability: How leaders with a late-stage action logic design and engage in sustainability initiatives. *Dissertation Abstracts International*, UMI No. 3498378.
4. J. Ehrenfeld & A. Hoffman (2013) *Flourishing: A frank conversation about sustainability*. Stanford, CA: Stanford University Press. p. xvii.

Chapter 1: Ecologically awake

5. T. Hemenway (2000) *Gaia's garden: A guide to home-scale permaculture*. White River Junction, VT: Chelsea Green Publishing.

6.   M. Pollan (2006) *The omnivore's dilemma: A natural history of four meals*. New York: Penguin.

7.   B. Kingsolver (2007) *Animal, vegetable, miracle: A year of food life*. New York: HarperCollins. p. 5.

8.   L. Brown (2010) *Plan B 4.0: Rescuing a planet under stress and a civilization in trouble*. New York: Norton.

9.   Oregon State University (2012) Outbreak of California Fivespined Ips Bark Beetle. Retrieved from http://oregonstate.edu/dept/mcarec/sites/default/files/outbreak_of_california_fivespined_ips_bark_beetle.pdf.

10. F. Capra (1996) *The web of life: A new synthesis of mind and matter*. London: HarperCollins.

11. The Bioneers organization has published several anthologies, including: K. Ausubel (1997) *Bioneers: A declaration of interdependence*. White River Junction, VT: Chelsea Green Publishing; and Z. Barlow & M. Stone (Eds.) (2005) *Ecological literacy: Educating our children for a sustainable world*. San Francisco: CA: Sierra Club Books.

12. R. Anderson (1998) *Midcourse correction: Toward a sustainable enterprise—the interface model*. White River Junction, VT: Chelsea Green Publishing.

13. Although many students and faculty were involved, particularly noteworthy were the contributions of sustainability student leaders Sean Franks and Tom Letchworth, and faculty members Dr Vincent Smith, Dr Mark Shibley, Dr Greg Jones and Byrone Marlowe. See https://sou.edu/sustainable/center-for-sustainability/index.html.

Chapter 2: The limits of "sustainability"

14. Currently, there is a new Director at the recently combined SOU School of Business, Communication, and Environment with a strong background in environmental science and climatology.

15. T. Mohin (2012) *Changing business from the inside out: A treehugger's guide to working in corporations*. Sheffield, UK: Greenleaf Publishing. p. 224.

16. J. Ehrenfeld & A. Hoffman (2013) *Flourishing: A frank conversation about sustainability*. Stanford, CA: Stanford University Press. p. 17.

17.   A. Drengson & Y. Inoue (Eds.) (1995) *The Deep Ecology Movement: An introductory anthology.* Berkeley, CA: North Atlantic Books.
18.   T. Roszak, M. Gomes & A. Kanner (Eds.) (1995) *Ecopsychology: Restoring the Earth, healing the mind.* Berkeley, CA: Sierra Club Books.

Chapter 3: Perspectives on ecological worldviews

19.   M.A. Hart (2010) Indigenous worldviews, knowledge, and research: The development of an indigenous research paradigm. *Journal of Indigenous Voices in Social Work*, 1, 1-16.
20.   M. Koltko-Rivera (2004) The psychology of worldviews, *Review of General Psychology*, 8(1), 3-58.
21.   P. Ray & S. Anderson (2000) *The cultural creatives: How 50 million people are changing the world.* New York: Three Rivers Press.
22.   W. Overton (2006) Developmental psychology: Philosophy, concepts, methodology. In W. Damon & R.J. Lerner (Eds.). *Handbook of child psychology, vol. 1* (pp. 18-54). Hoboken, NJ: John Wiley & Sons.
23.   B.C. Brown (2012) Conscious leadership for sustainability: How leaders with a late-stage action logic design and engage in sustainability initiatives. *Dissertation Abstracts International*, UMI No. 3498378.
24.   S. Esbjörn-Hargens & M. Zimmerman (2009) *Integral ecology: Uniting multiple perspectives on the natural world.* Boston, MA: Integral Books.
25.   L. White (1967) The historical roots of our ecologic crisis. *Science*, 155, 1203-1207.
26.   A. Naess, A. Drengson & B. Devall (2008) *The ecology of wisdom: Writings by Arne Naess.* Berkeley, CA: Counterpoint Press.
27.   T. Roszak, M. Gomes & A. Kanner (Eds.) (1995) *Ecopsychology: Restoring the Earth, healing the mind.* Berkeley, CA: Sierra Club Books.
28.   T. Doherty (2012) Psychology as if the whole earth mattered. *Oregon State Bar Sustainable Future: The Long View*, 9 (Spring 2012), 7-8.

29. T.S. Kuhn (1962) *The structure of scientific revolutions*. Chicago, IL: University of Chicago Press.

30. D. Meadows, J. Randers & D. Meadows (2004) *Limits to growth: The 30-year update*. White River Junction, VT: Chelsea Green Publishing.

31. R. Dunlap (2008) The New Environmental Paradigm Scale: From marginality to worldwide use. *The Journal of Environmental Education*, 40(1), 3-18.

32. A. Hedlund-de Witt (2012) Exploring worldviews and their relationships to sustainable lifestyles: Towards a new conceptual and methodological approach. *Ecological Economics*, 84, 74-83.

33. H. Daly (1996) *Beyond growth: The economics of sustainable development*. Boston, MA: Beacon.

34. R. Beddoe, R. Costanza, J. Farley, E. Garza, J. Kent, I. Kubiszewski, ... J. Woodward (2009) Overcoming systemic roadblocks to sustainability: The evolutionary redesign of worldviews, institutions, and technologies. *National Academy of Science*, 106(8), 2483-2489.

35. R. Dietz & D. O'Neill (2013) *Enough is enough: Building a sustainable economy in a world of finite resources*. San Francisco, CA: Berrett Koehler. p. 34.

36. Four Arrows (aka D. Jacobs) (2013) *Teaching truly: A curriculum to indigenize mainstream education*. New York: Peter Lange.

37. Four Arrows (aka D. Jacobs) (2008) *The authentic dissertation: Alternative ways of knowing, research, and representation*. New York: Routledge.

38. D. Abram (2010) *Becoming animal: An earthly cosmology*. New York: Random House.

39. D. Abram (1996) *The spell of the sensuous*. New York: Vintage Books. p. 20.

40. A. Hoffman (2015) The cultural schism of climate change: How Science takes a back seat to identity politics in the US. Retrieved from http://www.corporateecoforum.com/cultural-schism-climate-change-science-takes-back-seat-identity-politics-u-s/.

Chapter 4: Life experiences that shape ecological worldviews

41. P. Hawken (1994) *The ecology of commerce: A declaration of sustainability.* New York: Harper Business.

Chapter 5: Anthropocentric blindness

42. J. Hillman (1995) A psyche the size of the Earth: A psychological forward. In T. Roszak, M. Gomes & A. Kanner (Eds.), *Ecopsychology: Restoring the Earth, healing the mind* (pp. xvii-xxiii). Berkeley, CA: Sierra Club Books.
43. L. White (1967) The historical roots of our ecologic crisis. *Science,* 155, 1203-1207.
44. M. Tercek & J. Adams (2013) *Nature's fortune: How business and society thrived by investing in nature.* Philadelphia, PA: Basic Books.
45. M. Atwood (2011) What is our proper relationship with nature? *Resurgence and Ecologist,* 268, 4-32.
46. D. Goleman (2009) *Ecological intelligence: The hidden impacts of what we buy.* New York: Broadway Books.
47. D. Goleman, L. Bennett & Z. Barlow (2012) *Eco-literate: How educators are cultivating emotional, social, and ecological intelligence.* San Francisco, CA: Jossey-Bass.
48. F.M. Lappe (2011) *EcoMind: Changing the way we think, to create the world we want.* New York: Nations Books. p. 16.
49. L. Gray (1995) Shamanic counseling and ecopsychology. In T. Roszak, M. Gomes & A. Kanner (Eds.), *Ecopsychology: Restoring the Earth, healing the mind* (pp. 172-182). Berkeley, CA: Sierra Club Books. p. 182.
50. Quoted in Four Arrows (aka D. Jacobs) (2006) *Unlearning the language of conquest: Scholars challenge anti-Indianism in America.* Austin, TX: University of Texas Press. p. 260.
51. Quoted in Four Arrows (aka D. Jacobs) (2008) *The authentic dissertation: Alternative ways of knowing, research, and representation.* New York: Routledge. p. 20.

Chapter 6: The ecological self

52. A. Naess (1995) Self-realization: An ecological approach to being in the world. In A. Drengson & Y. Inoue (Eds.), *The Deep Ecology Movement: An introductory anthology* (pp. 13-30). Berkeley, CA: North Atlantic Books.

53. P. Shepard quoted in B. Devall (1995) The ecological self. In A. Drengson & Y. Inoue (Eds.), *The Deep Ecology Movement: An introductory anthology* (pp. 101-123). Berkeley, CA: North Atlantic Books. p. 102.

54. B. Devall & G. Sessions (1985) *Deep ecology: Living as if nature mattered.* Layton, UT: Gibbs Smith.

55. B. Devall (1995) The ecological self. In A. Drengson & Y. Inoue (Eds.), *The Deep Ecology Movement: An introductory anthology* (pp. 101-123). Berkeley, CA: North Atlantic Books.

56. L. Sewall (1995) The skill of ecological perception. In T. Roszak, M. Gomes & A. Kanner (Eds.), *Ecopsychology: Restoring the Earth, healing the mind* (pp. 201-215). Berkeley, CA: Sierra Club Books.

57. E. Bragg (1996) Towards ecological self: Deep ecology meets constructionist self-theory. *Journal of Environmental Psychology,* 16, 93-108.

58. S. Esbjörn-Hargens & M. Zimmerman (2009) *Integral ecology: Uniting multiple perspectives on the natural world.* Boston, MA: Integral Books.

59. K. Rogers (2012) Exploring our ecological selves within learning organizations. *The Learning Organization,* 19(1), 28-37.

60. P. Kahn (1999) *The human relationship with nature: Development and culture.* Boston, MA: MIT Press.

61. R. Louv (2008) *Last child in the woods: Saving our children from nature-deficit disorder.* Chapel Hill, NC: Algonquin Books.

62. J. Macy & M. Brown (2014) *Coming back to life: The updated guide to the work that reconnects.* Gabriola Island, BC: New Society Publishers.

63. J. Seed, J. Macy, P. Fleming & A. Naess (1988) *Thinking like a mountain: Towards a council of all beings.* Philadelphia, PA: New Society.

64. J. Macy & M. Brown (2014) *Coming back to life: The updated guide to the work that reconnects.* Gabriola Island, BC: New Society Publishers.

Chapter 8: Interior dimensions of leadership

65. P.G. Northouse (2013) *Leadership: Theory and practice.* Thousand Oaks, CA: Sage.
66. R. Greenleaf (1977) *Servant leadership: A journey into the nature of legitimate power and greatness.* Mahwah, NJ: Paulist Press.
67. D. Goleman, R. Boyatzis & A. McKee (2002) *Primal leadership: Realizing the power of emotional intelligence.* Boston, MA: Harvard Business School Press.
68. D. Goleman, R. Boyatzis & A. McKee (2002) *Primal leadership: Realizing the power of emotional intelligence.* Boston, MA: Harvard Business School Press. p. 5.
69. P. Senge, R. Ross, B. Smith, C. Roberts & A. Kleiner (1994) *The fifth discipline fieldbook: Strategies and tools for building a learning organization.* New York: Doubleday.
70. P.G. Northouse (2013) *Leadership: Theory and practice.* Thousand Oaks, CA: Sage.

Chapter 9: Human development

71. J. Piaget (1954) *The construction of reality in the child.* New York: Basic Books.
72. For example, L. Kohlberg (1969) Stage and sequence: The cognitive developmental approach to socialization. In D. Goslin (Ed.), *Handbook of socialization: Theory and research* (pp. 347-480). New York: Rand McNally.
73. S.R. Cook-Greuter (2004) Making the case for a developmental perspective. *Industrial and Commercial Training*, 36(6/7), 275-281.
74. W. Torbert (2004) *Developmental action inquiry.* San Francisco, CA: Berrett Koehler.
75. B.C. Brown (2013) Communicating sustainability to different worldviews. Retrieved from http://www.cruxcatalyst.com/2013/08/15/communicating-sustainability-to-different-worldviews/.

76. See http://www.williamrtorbert.com/about-bill-torbert/
    action-inquiry-associates-team/.

77. D. Rooke & W.R. Torbert (2005) Seven transformations of
    leadership, *Harvard Business Review*, 83, 66-76. p. 67.

78. For example, W.R. Torbert & E. Herdman-Barker (2008) Generating
    and measuring practical differences in leadership performance at
    postconventional action-logics: Developing the Harthill Leadership
    Development Profile. In A. Coombs, A. Pfaffenberger & P. Marko
    (Eds.), *The postconventional personality: Perspectives on higher
    development* (pp. 39-56). New York: SUNY Press. D. Rooke & W.R.
    Torbert (1998) Organizational transformation as a function of
    CEO's developmental stage, *Organization Development Journal*,
    16(1), 11-28.

79. R. Kegan (1994) *In over our heads: The mental demands of modern
    life*. Cambridge, MA: Harvard University Press. p. 78.

80. J. Rowson (2015) Bob's big idea: why we are living longer. Retrieved
    from https://www.thersa.org/discover/publications-and-articles/
    rsa-blogs/2015/01/bobs-big-idea-why-we-are-living-longer/.

81. O. Boiral, M. Cayer & C.M. Baron (2009) The action logics of
    environmental leadership: A developmental perspective. *Journal of
    Business Ethics*, 85, 479-499.

82. W. Visser & A. Crane (2010) Corporate sustainability and the
    individual: Understanding what drives sustainability professionals as
    change agents. *SSRN Working Paper Series*, 25 February 2010.

83. A. Lynam (2012) Navigating a geography of sustainability
    worldviews: A developmental map. *Journal of Sustainability
    Education*, 3, 1-14.

84. S. Divecha & B.C. Brown (2013) Integral sustainability: Correlating
    actions logics with sustainability. *Journal of Integral Theory and
    Practice*, 8(3/4), 197-211.

85. K. Rogers (2012) Exploring our ecological selves within learning
    organizations. *The Learning Organization*, 19(1), 28-37.

86. I. Rimanoczy (2013) *Big bang being: Developing the sustainability
    mindset*. Sheffield, UK: Greenleaf Publishing.

87. R. Anderson (1998) *Midcourse correction: Toward a sustainable enterprise—the interface model*. White River Junction, VT: Chelsea Green Publishing. p. 40.
88. B.C. Brown (2012) Conscious leadership for sustainability: How leaders with a late-stage action logic design and engage in sustainability initiatives. *Dissertation Abstracts International*, UMI No. 3498378.

Chapter 10: Expressions of post-conventional worldviews

89. W. Torbert (2004) *Developmental action inquiry*. San Francisco, CA: Berrett Koehler.
90. W.R. Torbert (2010) Listening into the dark: An essay testing the validity and efficacy of developmental action inquiry for describing and encouraging the transformation of self, society, and scientific inquiry. Paper presented at the Symposium on Research Across Boundaries, Luxembourg.
91. K. Wilber (2000) *Integral psychology: Consciousness, spirit, psychology, therapy*. Boston, MA: Shambala.

Chapter 11: The collaborator-in-chief (with an ecological worldview)

92. P. Senge, B. Smith, N. Kruschwitz, J. Laur & S. Schley (2008) *The necessary revolution: How individuals and organizations are working together to create a sustainable world*. New York: Doubleday. p. 233.
93. E. Maltz & S. Schein (2013) Cultivating shared value initiatives: A three c's approach, *Journal of Corporate Citizenship*, 47, 55-74.
94. L. Hill (2014) How to manage for collective creativity. Retrieved from http://www.ted.com/talks/linda_hill_how_to_manage_for_collective_creativity.

Chapter 12: Cultivating a new psychology for sustainability leadership

95. R. Vercoe & R. Brinkmann (2012) A tale of two sustainabilities: Comparing sustainability in the global north and south to uncover meaning for educators. Retrieved from http://www.jsedimensions. org/wordpress/content/a-tale-of-two-sustainabilities-comparing-sustainability-in-the-global-north-and-south-to-uncover-meaning-for-educators_2012_03/.

96. D. Goleman, L. Bennett & Z. Barlow (2012) *Eco-Literate: How educators are cultivating emotional, social, and ecological intelligence.* San Francisco, CA: Jossey-Bass.

97. J. Daniel (2005) *Rogue river journal.* Berkeley, CA: Counterpoint Books.

98. B. Devall (1995) The ecological self. In A. Drengson & Y. Inoue (Eds.), *The Deep Ecology Movement: An introductory anthology* (pp. 101-123). Berkeley, CA: North Atlantic Books.

99. J. Macy & M. Brown (2014) *Coming back to life: The updated guide to the work that reconnects.* Gabriola Island, BC: New Society Publishers.

100. L. Sewall (1995) This skill of ecological perception. In T. Roszak, M. Gomes & A. Kanner (Eds.), *Ecopsychology: Restoring the Earth, healing the mind* (pp. 201-215). Berkeley, CA: Sierra Club Books.

101. S. Esbjörn-Hargens & M. Zimmerman (2009) *Integral ecology: Uniting multiple perspectives on the natural world.* Boston, MA: Integral Books.

102. W. Torbert (2004) *Action inquiry: The secret of timely and transforming leadership.* San Francisco, CA: Berrett Koehler.

103. R. Heifetz, A. Grashow & M. Linsky (2009) *The practice of adaptive leadership: Tools and tactics for changing your organization and the world.* Cambridge, MA: Harvard Business School Publishing.

104. B. Joiner & S. Josephs (2007) *Leadership agility: Five levels of mastery for anticipating and initiating change.* San Francisco, CA: Jossey-Bass.

105. R. Kegan & L. Lahey (2009) *Immunity to change: How to overcome it and unlock the potential in yourself and your organizations.* Cambridge, MA: Harvard University Press.

106. D. Rooke & W.R. Torbert (2005) Seven transformations of leadership, *Harvard Business Review*, 83, 66-76.

107. D. Kiron, N. Kruschwitz, H. Rubel, M. Reeves & S. Fuisz-Kehrbach (2013) *Sustainability's next frontier: Walking the talk on the sustainability issues that matter most.* Cambridge, MA: MIT Sloan Management Review Research Report. p. 3.

Chapter 13: Multinational executives as human trim tabs

108. See http://bfi.org.

109. See http://www.jeanhouston.org.

110. See http://www.ceres.org/bicep.

111. J. Makower (2012) Two steps forward: Why aren't there more Ray Andersons? Retrieved from http://www.greenbiz.com/blog/2012/08/06/why-aren't-there-more-ray-andersons.

112. See https://netimpact.org/conference/live.

113. M. Gunther (2014) Starbucks CEO Howard Schultz: Retail employees have some of the worst pay and benefits of any industry. Retrieved from http://www.theguardian.com/sustainable-business/2014/nov/12/starbucks-ceo-howard-schultz-retail-employees-have-some-of-the-worst-pay-and-benefits-of-any-industry.

Appendix A: Ecological Sustainability Worldview Assessment Tool (E-SWAT)

114. F.M. Lappe (2011) *EcoMind: Changing the way we think, to create the world we want.* New York: Nations Books. p. 3.

115. J. Campbell (2008) *The hero with a thousand faces* (3rd ed.) Novato, CA: New World Library.

116. See http://mankindproject.org.

117. P. Senge (1990) *The fifth discipline: The art and practice of the learning organization.* New York: Bantam Doubleday Dell.

Appendix B: Research methodology and description of participants

118. J. Creswell (2009) *Research design: Qualitative, quantitative, and mixed methods approaches*. Thousand Oaks, CA: Sage.

119. Y.S. Lincoln & E.G. Guba (1985) *Naturalistic inquiry*. Newbury Park, CA: Sage.

120. E. Babbie (2002) *The basics of social research*. Belmont, CA: Wadsworth Thomson.

121. V. Braun & V. Clarke (2006) Using thematic analysis in psychology. *Qualitative Research in Psychology*, 3, 77-101.

122. V. Bentz & J. Shapiro (1998) *Mindful inquiry in social research*. Thousand Oaks, CA: Sage.

123. K. Rogers (2013) *Hermeneutic methods for a globalized world*. Manuscript submitted for publication.

References

Abram, D. (1996) *The spell of the sensuous*. New York: Vintage Books.
—— (2010) *Becoming animal: An earthly cosmology*. New York: Random House.
Anderson, R. (1998) *Midcourse correction: Toward a sustainable enterprise—the interface model*. White River Junction, VT: Chelsea Green Publishing.
Atwood, M. (2011) What is our proper relationship with nature? *Resurgence and Ecologist*, 268, 4-32.
Ausubel, K. (1997) *Bioneers: A declaration of interdependence*. White River Junction, VT: Chelsea Green Publishing.
Babbie, E. (2002) *The basics of social research*. Belmont, CA: Wadsworth Thomson.
Barlow, Z., & Stone, M. (Eds.) (2005) *Ecological literacy: Educating our children for a sustainable world*. San Francisco, CA: Sierra Club Books.
Bateson, G. (1972) *Steps to an ecology of mind*. Chicago, IL: HarperCollins.
Beddoe, R., Costanza, R., Farley, J., Garza, E., Kent, J., Kubiszewski, I., … Woodward, J. (2009) Overcoming systemic roadblocks to sustainability: The evolutionary redesign of worldviews, institutions, and technologies. *National Academy of Science*, 106(8), 2483-2489.

Bentz, V., & Shapiro, J. (1998) *Mindful inquiry in social research.*
    Thousand Oaks, CA: Sage.
Boiral, O., Cayer, M., & Baron, C.M. (2009) The action logics of
    environmental leadership: A developmental perspective. *Journal of
    Business Ethics,* 85, 479-499.
Bragg, E. (1996) Towards ecological self: Deep ecology meets
    constructionist self-theory. *Journal of Environmental Psychology,* 16,
    93-108.
Braun, V., & Clarke, V. (2006) Using thematic analysis in psychology.
    Qualitative Research in Psychology, 3, 77-101.
Brown, B.C. (2012) Conscious leadership for sustainability: How leaders
    with a late-stage action logic design and engage in sustainability
    initiatives. *Dissertation Abstracts International,* UMI No. 3498378.
—— (2013) Communicating sustainability to different worldviews.
    Retrieved from http://www.cruxcatalyst.com/2013/08/15/
    communicating-sustainability-to-different-worldviews/.
Brown, L. (1995) Ecopsychology and the environmental revolution:
    An environmental foreword. In T. Roszak, M. Gomes & A. Kanner
    (Eds.), *Ecopsychology: Restoring the Earth, healing the mind*
    (pp. xiii-xvi). Berkeley, CA: Sierra Club Books.
—— (2010) *Plan B 4.0: Rescuing a planet under stress and a civilization
    in trouble.* New York: Norton.
Campbell, J. (2008) *The hero with a thousand faces* (3rd ed.) Novato,
    CA: New World Library.
Capra, F. (1996) *The web of life: A new synthesis of mind and matter.*
    London: HarperCollins.
Cook-Greuter, S.R. (2000) Mature ego development: A gateway to ego
    transcendence? *Journal of Adult Development,* 7(4), 227-240.
—— (2004) Making the case for a developmental perspective. *Industrial
    and Commercial Training,* 36(6/7), 275-281.
Creswell, J. (2009) *Research design: Qualitative, quantitative, and mixed
    methods approaches.* Thousand Oaks, CA: Sage.
Daly, H. (1996) *Beyond growth: The economics of sustainable
    development.* Boston, MA: Beacon.
Daniel, J. (2005) *Rogue river journal.* Berkeley, CA: Counterpoint Books.

Devall, B. (1995) The ecological self. In A. Drengson & Y. Inoue (Eds.), *The Deep Ecology Movement: An introductory anthology* (pp. 101-123). Berkeley, CA: North Atlantic Books.

Devall, B., & Sessions, G. (1985) *Deep ecology: Living as if nature mattered.* Layton, UT: Gibbs Smith.

Dietz, R., & O'Neill, D. (2013) *Enough is enough: Building a sustainable economy in a world of finite resources.* San Francisco, CA: Berrett Koehler.

Divecha, S., & Brown, B.C. (2013) Integral sustainability: Correlating actions logics with sustainability. *Journal of Integral Theory and Practice,* 8(3/4), 197-211.

Doherty, T. (2012) Psychology as if the whole earth mattered. *Oregon State Bar Sustainable Future: The Long View,* 9 (Spring 2012), 7-8.

Drengson, A., & Inoue, Y. (Eds.) (1995) *The Deep Ecology Movement: An introductory anthology.* Berkeley, CA: North Atlantic Books.

Dunlap, R. (2008) The New Environmental Paradigm Scale: From marginality to worldwide use. *The Journal of Environmental Education,* 40(1), 3-18.

Dunlap, R.E., & Van Liere, K.D. (1978) The "New Environmental Paradigm": A proposed measuring instrument and preliminary results. *The Journal of Environmental Education,* 9(4), 10-19.

Ehrenfeld, J., & Hoffman, A. (2013) *Flourishing: A frank conversation about sustainability.* Stanford, CA: Stanford University Press.

Esbjörn-Hargens, S., & Zimmerman, M. (2009) *Integral ecology: Uniting multiple perspectives on the natural world.* Boston, MA: Integral Books.

Ferdig, M. (2007) Sustainability leadership: Co-creating a sustainable future. *Journal of Change Management,* 1(7), 25-35.

Fischer, A. (2013) *Radical eco-psychology: Psychology in the service of life.* Albany, NY: SUNY Press.

Four Arrows (aka D. Jacobs) (2006) *Unlearning the language of conquest: Scholars challenge anti-Indianism in America.* Austin, TX: University of Texas Press.

—— (2008) *The authentic dissertation: Alternative ways of knowing, research, and representation.* New York: Routledge.

—— (2013) *Teaching truly: A curriculum to indigenize mainstream education*. New York: Peter Lange.

Goleman, D. (1998) *Working with emotional intelligence*. New York: Bantam.

—— (2009) *Ecological intelligence: The hidden impacts of what we buy*. New York: Broadway Books.

Goleman, D., Bennett, L., & Barlow, Z. (2012) *Eco-literate: How educators are cultivating emotional, social, and ecological intelligence*. San Francisco, CA: Jossey-Bass.

Goleman, D., Boyatzis, R., & McKee, A. (2002) *Primal leadership: Realizing the power of emotional intelligence*. Boston, MA: Harvard Business School Press.

Greenleaf, R. (1977) *Servant leadership: A journey into the nature of legitimate power and greatness*. Mahwah, NJ: Paulist Press.

Gray, L. (1995) Shamanic counseling and ecopsychology. In T. Roszak, M. Gomes & A. Kanner (Eds.), *Ecopsychology: Restoring the Earth, healing the mind* (pp. 172-182). Berkeley, CA: Sierra Club Books.

Grey, W. (1993), Anthropocentrism and deep ecology. *Australasian Journal of Philosophy*, 71(4), 463-475.

Gunther, M. (2014) Starbucks CEO Howard Schultz: Retail employees have some of the worst pay and benefits of any industry. Retrieved from http://www.theguardian.com/sustainable-business/2014/nov/12/starbucks-ceo-howard-schultz-retail-employees-have-some-of-the-worst-pay-and-benefits-of-any-industry.

Hart, M.A. (2010) Indigenous worldviews, knowledge, and research: The development of an indigenous research paradigm. *Journal of Indigenous Voices in Social Work*, 1, 1-16.

Hawken, P. (1994) *The ecology of commerce: A declaration of sustainability*. New York: Harper Business.

—— (2008) *Blessed unrest: How the largest movement in the world came into being and why no one saw it coming*. New York: Viking.

Hawken, P., Lovins, A., & Lovins, H. (1999) *Natural capitalism: Creating the next industrial revolution*. New York: Little, Brown.

Hedlund-de Witt, A. (2012) Exploring worldviews and their relationships to sustainable lifestyles: Towards a new conceptual and methodological approach. *Ecological Economics*, 84, 74-83.

Heifetz, R., Grashow, A., & Linsky, M. (2009) *The practice of adaptive leadership: Tools and tactics for changing your organization and the world.* Cambridge, MA: Harvard Business School Publishing.

Hemenway, T. (2000) *Gaia's garden: A guide to home-scale permaculture.* White River Junction, VT: Chelsea Green Publishing.

Hill, L. (2014) How to manage for collective creativity. Retrieved from http://www.ted.com/talks/linda_hill_how_to_manage_for_collective_creativity.

Hillman, J. (1995) A psyche the size of the Earth: A psychological forward. In T. Roszak, M. Gomes & A. Kanner (Eds.), *Ecopsychology: Restoring the Earth, healing the mind* (pp. xvii-xxiii). Berkeley, CA: Sierra Club Books.

Hoffman, A. (2015) The cultural schism of climate change: How Science takes a back seat to identity politics in the US. Retrieved from http://www.corporateecoforum.com/cultural-schism-climate-change-science-takes-back-seat-identity-politics-u-s/.

Hulme, M. (2009) *Why we disagree about climate change.* New York: Cambridge University Press.

Jacobs, D. (1997) *Primal awareness: A true story of survival, transformation and awakening with the Raramuri shamans of Mexico.* Rochester, VT: Inner Traditions International.

Joiner, B., & Josephs, S. (2007) *Leadership agility: Five levels of mastery for anticipating and initiating change.* San Francisco, CA: Jossey-Bass.

Kahn, P. (1999) *The human relationship with nature: Development and culture.* Boston, MA: MIT Press.

—— (2011) *Technological nature: Adaptation and the future of human life.* Cambridge, MA: MIT Press.

Kahn, P., & Hasbach, P. (2012) *Eco-psychology: Science, totems, and the technological species.* Cambridge, MA: MIT Press.

Kegan, R. (1994) *In over our heads: The mental demands of modern life.* Cambridge, MA: Harvard University Press.

Kegan, R., & Lahey, L. (2009) *Immunity to change: How to overcome it and unlock the potential in yourself and your organizations.* Cambridge, MA: Harvard University Press.

Kingsolver, B. (2007) *Animal, vegetable, miracle: A year of food life.* New York: HarperCollins.

Kiron, D., Kruschwitz, N., Rubel, H., Reeves, M., & Fuisz-Kehrbach, S. (2013) *Sustainability's next frontier: Walking the talk on the sustainability issues that matter most*. Cambridge, MA: MIT Sloan Management Review Research Report.

Kohlberg, L. (1969) Stage and sequence: The cognitive developmental approach to socialization. In D. Goslin (Ed.), *Handbook of socialization: Theory and research* (pp. 347-480). New York: Rand McNally.

Koltko-Rivera, M. (2004) The psychology of worldviews. *Review of General Psychology*, 8(1), 3-58.

Kuhn, T.S. (1962) *The structure of scientific revolutions*. Chicago, IL: University of Chicago Press.

Lappe, F.M. (2011) *EcoMind: Changing the way we think, to create the world we want*. New York: Nations Books.

Lincoln, Y.S., & Guba, E.G. (1985) *Naturalistic inquiry*. Newbury Park, CA: Sage.

Louv, R. (2008) *Last child in the woods: Saving our children from nature-deficit disorder*. Chapel Hill, NC: Algonquin Books.

Lynam, A. (2012) Navigating a geography of sustainability worldviews: A developmental map. *Journal of Sustainability Education*, 3, 1-14.

Macy, J. (1989) Awakening to the ecological self. In J. Plant (Ed.), *Healing the wounds: The promise of eco-feminism* (pp. 201-211). Philadelphia, PA: New Society Publishers.

—— (2007) *World as lover, world as self: Courage for global justice and ecological Renewal*. Berkeley, CA: Parallax Press.

Macy, J., & Brown, M. (2014) *Coming back to life: The updated guide to the work that reconnects*. Gabriola Island, BC: New Society Publishers.

Makower, J. (2012) Two steps forward: Why aren't there more Ray Andersons? Retrieved from http://www.greenbiz.com/blog/2012/08/06/why-aren't-there-more-ray-andersons.

Maltz, E., & Schein, S. (2013) Cultivating shared value initiatives: A three c's approach, *Journal of Corporate Citizenship*, 47, 55-74.

McDonough, W., & Braungart, M. (2002) *Cradle to cradle: Remaking the way we make things*. New York: North Point Press.

McKibben, B. (2010) *Eaarth: Making a life on a tough new planet*. New York: Times Books.

Meadows, D.H. (2008) *Thinking in systems: A primer*. White River Junction, VT: Chelsea Green.

Meadows, D., Randers, J., & Meadows, D. (2004) *Limits to growth: The 30-year update*. White River Junction, VT: Chelsea Green Publishing.

Mohin, T. (2012) *Changing business from the inside out: A treehugger's guide to working in corporations*. Sheffield, UK: Greenleaf Publishing.

Naess, A. (1995) Self-realization: An ecological approach to being in the world. In A. Drengson & Y. Inoue (Eds.), *The Deep Ecology Movement: An introductory anthology* (pp. 13-30). Berkeley, CA: North Atlantic Books.

Naess, A., Drengson, A., & Devall, B. (2008) *The ecology of wisdom: Writings by Arne Naess*. Berkeley, CA: Counterpoint Press.

Northouse, P.G. (2013) *Leadership: Theory and practice*. Thousand Oaks, CA: Sage.

Oregon State University (2012) Outbreak of California Fivespined Ips Bark Beetle. Retrieved from http://oregonstate.edu/dept/mcarec/sites/default/files/outbreak_of_california_fivespined_ips_bark_beetle.pdf.

Overton, W. (2006) Developmental psychology: Philosophy, concepts, methodology. In W. Damon & R.J. Lerner (Eds.), *Handbook of child psychology, vol. 1* (pp. 18-54). Hoboken, NJ: John Wiley & Sons.

Piaget, J. (1954) *The construction of reality in the child*. New York: Basic Books.

Pollan, M. (2006) *The omnivore's dilemma: A natural history of four meals*. New York: Penguin.

Ray, P., & Anderson, S. (2000) *The cultural creatives: How 50 million people are changing the world*. New York: Three Rivers Press.

Rimanoczy, I. (2013) *Big bang being: Developing the sustainability mindset*. Sheffield, UK: Greenleaf Publishing.

Rogers, K. (2012) Exploring our ecological selves within learning organizations. *The Learning Organization*, 19(1), 28-37.

—— (2013) *Hermeneutic methods for a globalized world*. Manuscript submitted for publication.

Rooke, D., & Torbert, W.R. (1998) Organizational transformation as a function of CEO's developmental stage, *Organization Development Journal*, 16(1), 11-28.

—— (2005) Seven transformations of leadership, *Harvard Business Review*, 83, 66-76.

Roszak, T., Gomes, M., & Kanner, A. (Eds.) (1995) *Ecopsychology: Restoring the Earth, healing the mind*. Berkeley, CA: Sierra Club Books.

Rowson, J. (2015) Bob's big idea: why we are living longer. Retrieved from https://www.thersa.org/discover/publications-and-articles/rsa-blogs/2015/01/bobs-big-idea-why-we-are-living-longer/.

Seed, J., Macy, J., Fleming, P., & Naess, A. (1988) *Thinking like a mountain: Towards a council of all beings*. Philadelphia, PA: New Society.

Senge, P. (1990) *The fifth discipline: The art and practice of the learning organization*. New York: Bantam Doubleday Dell.

Senge, P., Ross, R., Smith, B., Roberts, C., & Kleiner, A. (1994) *The fifth discipline fieldbook: Strategies and tools for building a learning organization*. New York: Doubleday.

Senge, P., Smith, B., Kruschwitz, N., Laur, J., & Schley, S. (2008) *The necessary revolution: How individuals and organizations are working together to create a sustainable world*. New York: Doubleday.

Sewall, L. (1995) The skill of ecological perception. In T. Roszak, M. Gomes & A. Kanner (Eds.), *Ecopsychology: Restoring the Earth, healing the mind* (pp. 201-215). Berkeley, CA: Sierra Club Books.

Shepard, P. (1995) Nature and madness. In T. Roszak, M. Gomes & A. Kanner (Eds.), *Ecopsychology: Restoring the Earth, healing the mind* (pp. 21-40). Berkeley, CA: Sierra Club Books.

Tercek, M., & Adams, J. (2013) *Nature's fortune: How business and society thrived by investing in nature*. Philadelphia, PA: Basic Books.

Torbert, W.R. (2004) *Action inquiry: The secret of timely and transforming leadership*. San Francisco, CA: Berrett Koehler.

—— (2010) Listening into the dark: An essay testing the validity and efficacy of developmental action inquiry for describing and encouraging the transformation of self, society, and scientific inquiry.

Paper presented at the Symposium on Research Across Boundaries, Luxembourg.

Torbert, W. R., & Herdman-Barker, E. (2008) Generating and measuring practical differences in leadership performance at postconventional action-logics: Developing the Harthill Leadership Development Profile. In A. Coombs, A. Pfaffenberger & P. Marko (Eds.), *The postconventional personality: Perspectives on higher development* (pp. 39-56). New York: SUNY Press.

Vercoe, R., & Brinkmann, R. (2012) A tale of two sustainabilities: Comparing sustainability in the global north and south to uncover meaning for educators. Retrieved from http://www.jsedimensions. org/wordpress/content/a-tale-of-two-sustainabilities-comparing-sustainability-in-the-global-north-and-south-to-uncover-meaning-for-educators_2012_03/.

Visser, W., & Crane, A. (2010) Corporate sustainability and the individual: Understanding what drives sustainability professionals as change agents. *SSRN Working Paper Series*, 25 February 2010.

Weinreb, E. (2011) *CSO backstory: How chief sustainability officers reached the C-suite*. Retrieved from http://weinrebgroup.com/wp-content/uploads/2011/09/CSO-Back-Story-by-Weinreb-Group.pdf.

White, L. (1967) The historical roots of our ecologic crisis. *Science*, 155, 1203-1207.

Wilber, K. (2000) *Integral psychology: Consciousness, spirit, psychology, therapy*. Boston, MA: Shambala.

About the author

Dr Steve Schein is a sustainability
leadership educator, researcher and
organizational consultant. After 25 years
in the corporate world and a decade
in academia, he sees the evolution of
business leadership and education
towards ecological sustainability as
a global imperative. To that end, his
research focuses on the development of ecological and post-
conventional worldviews in the setting of multinational
corporate leadership.

His work has been published in the *Journal of Corporate
Citizenship*, the *Journal of Management of Global
Sustainability*, and presented at numerous conferences on
corporate responsibility. He is currently Affiliate Professor
of Sustainability Leadership at Southern Oregon University,
where he has been a faculty member since 2005 and founded

the certificate program in sustainability leadership in 2007.
Prior to joining the faculty at SOU, he was a certified public
accountant (CPA) and former CEO with senior management
experience in several companies.

His primary areas of consulting include senior leadership
development, cultivating mission-driven high-performance
cultures, tri-sector collaboration, and the integration of
environmental and socially responsible business practices.
He holds a PhD in Human Development and Organizational
Systems from Fielding Graduate University and a BA in
accounting from the University of Colorado. He currently
serves on the Board of Directors for Net Impact (https://
netimpact.org) and the GEOS Institute (http://www.
geosinstitute.org). He lives in Ashland, Oregon and can
be reached at steve@sustainableleadership.com or via his
website at www.steveschein.net.